The African Mobile Story

WIRELESS WORLD
RESEARCH FORUM®

RIVER PUBLISHERS SERIES IN COMMUNICATIONS

This series focuses on communications science and technology. This includes the theory and use of systems involving all terminals, computers, and information processors; wired and wireless networks; and network layouts, procontentsols, architectures, and implementations.

Furthermore, developments toward newmarket demands in systems, products, and technologies such as personal communications services, multimedia systems, enterprise networks, and optical communications systems.

- Wireless Communications
- Networks
- Security
- Antennas & Propagation
- Microwaves
- Software Defined Radio

For a list of other books in this series, visit www.riverpublishers.com

The African Mobile Story

Knud Erik Skouby

Professor, Director CMI
Aalborg University
& Chair WGA/ WWRF
Denmark

Idongesit Williams

Ph.d Fellow
Aalborg University
Denmark

WWRF Book Series Editor:
Knud Erik Skouby

River Publishers
Aalborg

Published, sold and distributed by:
River Publishers
Niels Jernes Vej 10
9220 Aalborg Ø
Denmark

ISBN: 978-87-93102-63-7 (Hard Copy)
978-87-93102-64-4 (Ebook)

Foreword

The WWRF Series in Mobile Telecommunications

The Wireless World Research Forum (WWRF) is a global organization bringing together researchers into a wide range of aspects of mobile and wireless communications, from industry and academia, to identify the key research challenges and opportunities. Members and meeting participants work together to present their research and develop white papers and other publications on the way to the Wireless World. Much more information on the Forum, and details of its publication programme, are available on the WWRF website www.wwrf.ch. The scope of WWRF includes not just the study of novel radio technologies and the development of the core network, but also the way in which applications and services are developed, and the investigation of how to meet user needs and requirements.

WWRF's publication programme includes use of social media, online publication via our website and special issues of well respected journals. In addition, where we have identified significant deserving subjects, WWRF is keen to support the publication of extended expositions of our material in book form, either singly-authored or bringing together contributions from a number of authors. This series, published by River Publications, is focused on treating important concepts in some depth and bringing them to a wide readership in a timely way. Some will be based on extending existing white papers, while others are based on the output from WWRF-sponsored events or from proposals from individual members.

I hope that each volume of this series will be useful and informative to its readership, and will also contribute to further debate and contributions to WWRF and more widely.

Dr Nigel Jefferies
WWRF Chairman

Contents

Preface

This book is inspired by ongoing work and activities in the World Research Wireless Forum (WWRF; www.wwrf.ch). The essence of the idea was to chronicle the development and growth of mobile telephony in Africa with a focus on Sub- Saharan Africa. Until the 1990's, aside from South Africa and in North Africa, the teledensity in Africa was very low. A remarkable growth in mobile has since then changed the picture in Sub-Saharan Africa completely. Today mobile telephony penetration in Africa is above 60% and still growing. Hence it seems important to understand what is driving this growth and what its impact might be on people's lives.

In a bid to understand the growth of mobile in Africa, views from authorities on the various facets of mobile development in Africa were sought. Authors from Nigeria, South Africa, Ghana, Denmark and Sweden have contributed stating different factors based on cases broadly covering the development of mobile in Africa. The cases covered cover regulation and policy; mobile service implementation and adoption, and the application of mobile in everyday life.

This book is a combination of analysis, discussion and narration. The basic story of this book has been the interrelationship between mobile market, mobile technology and telecom policies and regulations in Africa. Mobile telephony is discussed both from the historical point of view, the present development and challenges as well as the future perspective of mobile development.

The target of this book includes ICT regulators, Mobile Network Operators, International Telecom Agencies, Government ICT agencies, the research community, and academia.

Sincere thanks are due to authors Dr. Alison Gillwald, Executive Director of Research ICT Africa, Cape Town University; Enrico Calandro, Research fellow at Research ICT Africa; Dr. Godfrey Frempong, dep. Director of CSIR, Ghana; Idongesit Williams, Ph.D Fellow at CMI, AAU Copenhagen, Denmark; Benjamin Kwofie, Ph.D Fellow of CMI, AAU Copenhagen; Roslyn Layton consultant at Strand Consult, Denmark; Perpetual Crentsil, Post Doc.

researcher at University of Helsinki Finland; Dr. Nana Koffi Annan of Wisconsin University Ghana; Dr. G.O Ofori-Dwumfuo of Methodist University Ghana; Patrick Ohemeng Gyaase of CMI, AAU, Copenhagen; Dr. Kwaduo Owusu, of University of Ghana, Legon; Alejandro Lopez and Gail Krantzberg, both of Mcmasters University, Canada; Joseph Wilson, Nuhu Gapsiso and Musa Usman, all from University of Maiduguri; Nuhu Gapsiso; Abdullahi Isa, DMO Adjin and Kenneth Tsivor, Ph.D fellows, CMI, AAU Copenhagen. Also thanks to Prof. Anders Henten, CMI for assistance in the editing process and to Dr. Peter Tobbin for helping to select some authors.

We are grateful for inspiration and support from the publishers, ITU and WWRF.

Finally, but not least thanks to Idongesit Williams for his tireless effort in collecting the contributions and editing of the book.

Knud Erik Skouby,
Chair WGA/ WWRF
Professor, Director CMI/ Aalborg University
Copenhagen, January 2014

Introduction

By Knud Erik Skouby & Idongesit Williams

The focus of this book on the development of mobile communications in Africa is on Sub-Saharan Africa, focusing on the drivers and barriers for a development that is seen as capable of delivering powerful potential for broader societal development. Sub-Saharan Africa has during the past decade witnessed one of the fastest growing markets in mobile communication and this growth is recognized to have played a pivotal role in the socio-economic development of most countries. It has had a huge impact on residential living patterns; on business networks and models; and on government services and income sources. There are now more than half a billion mobile phones in use in Sub-Saharan Africa, representing one of the biggest dramatic surges in usage in mobile telecom's three-decade history.

It is estimated that the mobile industry in 2013 contributed over 6% of the region's gross domestic product (GDP), higher than any other comparable region globally and that the mobile ecosystem directly supports 3.3 million jobs and contributes $21 billion to public funding in the region (GSMA 2013).

This development has been enabled by different regulatory frameworks; private and public investments; standardization efforts; innovative business models leading to competitive markets and technology neutrality. Mobile expansion is therefore expected to have big benefits for continued economic growth in Sub-Saharan Africa, resulting in further socio-economic development. The expectation is also that the enablers will continue to create favorable relationships between policies, technology access, use and advancement in mobile.

The development of mobile telephony in Africa has not been without the help of market reforms. One of the important drivers has been liberalization. In an overview of the market reforms in the telecoms market and a brief assessment of the impact of these reforms on mobile broadband, South Africa and 12 other countries are used as cases for the assessment. The reforms have led to competitive markets which have in turn led to innovative business modeling which then led to innovative service delivery and consequently market growth fuelled by substantial investments. Economic growth has followed, with a steady expansion of gross domestic product (GDP) across the continent as companies and individuals become more efficient and more productive to a large extent enabled by information and communication technologies (ICT) – that especially in Africa to a large extent is based on mobile. The global trend where ICT plays a central role in the development of technology and in business ecosystems in the communications and media areas is very visible in the African development resulting in a shift from resource-based economies and growth to growth based on internal structural changes (see, e.g., UN 2013). An overview of the structural reforms and their general impact is given Chapters One to Three.

It is a basic understanding in the discussions of this book that the contribution of ICT is and has been essential in rearranging the modes of operation and business models and processes. The central and reconfiguring role of ICT is encapsulated in the terms e-health, e-learning, etc. The focus and aim of the sectors are generally not changed fundamentally; however, the use of ICT changes the work processes and also the processes of value creation and of control in the business processes. Mobile applications are encouraging greater productivity via better health as short messaging services (SMS) can spread information to help fight malaria, encourage child immunizations, and assist people living with diseases such as HIV/AIDS. Africa has proven a fertile developing ground for a wide range of mobile applications—the majority of which are SMS not browser-based. Although the availability of ICT and telecommunications services is certainly not the sole factor driving GDP growth, the growth enabling and restructuring impact of ICT in this broad sense is certainly substantial. Chapters Six to Twelve discuss sector impact of the mobile development and specific case studies.

If the fruitful relationships between industry, operators and governments hold, the African telecommunications success story is likely to continue. There are, however, several potential stumbling blocks on the way forward. Some are related to generally inconsistent and sometimes unpredictable regulatory environments including setting of license fees and regulations regarding pricing and service launch present constraints. There is frequently an absence of coordination across government agencies which all would benefit from network roll-out and coverage.. The case of energy supply and policy is discussed in Chapter Eleven. Other barriers are related to the market organization and dominant prepaid systems where a discussion is raised on how much lower the tariffs could go and how this can affect the ARPU of the mobile network operator, asdiscussed in Chapter Four. Further the general reliance on mobile terminals as access to the internet makes security threats on mobile telephony a special issue in Africa as discussed in Chapter Five. These potential stumbling blocks may cause investors to get concerned and take a hard look at future potentials of African versus other markets, e.g., blocking the way forward to 3G and 4G networks. Networks are still very much dominated by voice communications and text messaging in the form of GSM. 3G subscriptions still only represent 11% of the overall African market (ITU 2013).

This book aims at analyzing the current mobile scene in Africa and developments – their drivers and barriers – by providing overviews on regulations, markets and specific cases. The objective is to analyze and assess the impact of the fast growing mobile market on these different societal levels. The scope of impact assessment includes regulation/policies, markets with demand and supply dynamics, business models, innovative service delivery, standards and the role of mobile wireless in development.

In *Chapter 1* the cost of broadband across 13 African countries is assessed based on household and individual ICT data and use data comparing mobile and fixed access in different baskets. The chapter identifies the overall impact of mobile broadband and its resulting regulatory and public policy challenges. Areas where policy concerns are identified are, e.g., the overdependence on mobile broadband as compared to fixed broadband; the increase in the use of data services to make voice calls (eg skype); private paid-for SMS'ing vs free instant messaging; inadequate capacity for the growing need for international and national bandwidths; the challenge of broadband affordability vs broadband pricing. The last two issues mentioned point to the central role of public private interplay (PPI). In developing economies generally and in Africa specifically neither the state nor the private sector are on their own are able to meet broadband needs in increasingly information-dependent economies. It is concluded that the major policy challenge generally for Africa is to create the conditions for investment in backbone and backhaul networks.

Chapter 2 gives an overview of the mobile market liberalization in Africa with the most striking result being the rise in mobile network coverage in urban Africa from 16% to 90% in ten years. It includes a general overview and three specific cases where the mobile telephony penetration before and after telecom liberalization and the drivers that led to the growth of mobile telephony are discussed. The specific telecom market liberalization cases chosen are Ghana, Nigeria and Kenya based on the facts that

- Nigeria is the fastest growing mobile market in Africa.
- Ghana is one of the first countries to introduce mobile telephony in sub-Saharan Africa and they also have a high mobile telephony penetration rate.
- Kenya is one of the countries where the privatization effort transformed the incumbent into two competing companies with innovative mobile solutions including incorporation of Public Private Partnership in its regulatory effort.

The results of liberalization are viewed as positive and challenging as well. There are differences between the three case-countries – notably a low rural penetration in Kenya, but the overall development is similar. Liberalization of the telecommunications markets has generally led to the development of a competitive market in most African countries with reduction in mobile telephony tariff, competitive business models, and innovative mobile service delivery. The main challenges are development of mobile broadband and the possible need for another market reform mechanism to help push mobile into rural Africa.

Chapter 3 provides an overview of the developments and a status of telecommunications in a selection of African countries. There are brief overviews of regulatory trends; of technology trends and the current status of the main access technologies. Generally regulatory trends include liberalization, privatization and innovative licensing regimes with the aim of fostering development in the markets, but it is concluded that much still remains to be done to ensure regulatory proficiency. Technologically, the networks are dominated by 2G GSM and CDMA. The deployment of 3G is evolving in Africa, but mainly in the urban areas. Africa has in a sense kept pace with the developments in telecommunications technologies. In the mobile field, the newest technologies are relatively quickly implemented, but in a limited scale. In the fixed area there has been little development; fiber is used in the backbone and only to a very small extent in the access paths.

It is illustrated that access to traditional telecommunications services, such as voice telephony and SMS, but also to the internet, goes via mobile platforms. The underdeveloped state of telecommunications from the colonial era has meant that as the fixed access infrastructure is very scanty, the present and future of telecommunications in African countries will generally be based on mobile platforms.

Chapter 4 discusses the situation with prepaid dominating the markets in Africa. More than half of the world's mobile services are offered on a prepaid basis, however in most African countries, prepaid service makes up more than 90% of the market. The intention is to review the prepaid market and its challenges and suggest ways in which actors can improve the situation. Prepaid offers easy and convenient access to modern communication for all type of customers, but there some areas of the prepaid market in Africa that can be optimized and improved. A number of specific suggestions to improve the African prepaid model are listed including ways to induce customers to sign up for data services. Detailed statistics on the uptake of mobile broadband are provided in an annex. As broadband penetration increases, people tend to upgrade to a smartphone. Going forward, operators will likely price voice and SMS together to stave off the shift to a competitive service. The development of the prepaid market was helped by low cost terminals, particularly from Motorola and Nokia. Many of these feature phones are sophisticated enough to double as smartphones. There is no doubt that mobile is the hotbed of innovation in Africa and mobile penetration is high in Africa, so there is evidence that the market can work, but the mobile industry will need to become more efficient to deliver the broadband and mobile products in the future.

Chapter 5 addresses the situation that the shift from PC-based user interactions to mobile device user interactions has equally shifted the threat and security challenges to mobile devices and they are increasingly targets of malware threats because they are getting less secure with the move towards smartphones and the download of apps and thereby susceptible to end user manipulation through social engineering tactics. As the mobile phone is the primary access media to the Internet in Africa and Africans are very active on social media platforms, the problem is emerging as very serious in Africa. The spread of malware especially through android smartphones already has a long track record. A specified list of the different attacks and their targets is given emphasizing that the propagation of malware in social media is increasing in sophistication and strategies. Among the consequences are battery depletion, exposure of user privacy, financial loss and data leakage. There is a call for a solution involving mobile service providers in partnership with manufacturers, vendors, and service agents to create a system motivating and training users. If the security threats are not dealt with, serious setbacks to the growth and development of communication in Africa will arise.

Chapter 6 reviews and discusses a number of projects indicating showing that mobile phones have shown promise in providing greater access to healthcare to populations in African countries and especially in HIV treatment and prevention initiatives. There is widespread optimism that the concept of using ICTs to propel health and development will incontrovertibly lead to development. However, scaling up and sustaining these programs have less been addressed. Despite hundreds of m-Health pilot studies, there has been insufficient programmatic evidence to inform implementation, scale up m-Health solutions in HIV/AIDS initiatives. To understand a way forward and the role of new media as part of local hierarchies when a powerful new medium is appropriated, the specific communicative ecology related to HIV in Ghana, South Africa, and Uganda is analyzed using ethnographic data and review of publications. A number of suggestions for addressing challenges and constraints are presented. A key challenge in m-health is low availability of phones. Access to phones is still a problem in Africa despite the high penetration of the devices in the continent and it is a problem how various stakeholders in the mobile phone, service providers, government agencies, NGOs, and other industries can co-operate to provide applications in affordable and sustainable ways. This include a number of basic technological limitations that bear upon m-health communication initiatives. The underlying infrastructure—though not nearly as limited or expensive to deploy as some other ICTs like fixed line telephony or cable Internet—is not always universal. Patchy network coverage, service fluctuations, bandwidth limitations, and otherwise unreliable connectivity are abundant. However, it is concluded that mobile health interventions could be applied to a very broad range of health-related behaviors, but there is a need to assess interventions based on individual and community participation.

Chapter 7 looks at the use of mobile-ICT in education in Africa with focus on some cases in Ghana. It is a basic understanding in the chapter that one of the ways out for Africa in advancing education delivery is to take full advantage of the numerous opportunities offered by mobile computing and communication technologies. Further it is stated that Africa can be considered as having the highest mobile learning growth rate in the world with Nigeria being one of the main drivers of the growth rate witnessed across the continent. Overall trends in m-learning in some countries are discussed and details provided from pilots in Ghana. A positive and a negative main driver are identified: 1) the exponential growth of mobile phone subscriptions in the region, which is due to unprecedented development in mobile technologies and low costs for mobile devices and data plans. The liberalization and deregulation of the telecommunications sector has paved way for massive private investment in the mobile telecommunication industry. 2) Systemic failures in the delivery of traditional education.

Lack of ICT- skills and social cultural background of users are mentioned as the biggest barriers to the effective use mobile devices for teaching or learning purposes.

Chapter 8 is discussing the development where the traditional subsistence agriculture in Africa gradually is been replaced by market-oriented or commercial agriculture. Agricultural extension and advisory services play an important role in the transition. However, weak extension systems, have contributed to low or non-adoption of new agricultural technologies. An initiative in Ghana, Esoko Networks addresses this problem and meets the information needs of farmers and other players in the agricultural value chain by providing current market prices, matching bids and offers, weather forecasts, and news and other tips. Esoko is a mobile-enabled, cloud-based service operating also in 7 other African countries which could be used on a basic mobile phone or a computer with SMS and data connection functionality. The company's goal is to put more money into the hands of farmers by addressing the information asymmetry. that exists where farmers are frequently disadvantaged price-takers which often results in the farmers selling at a loss. Unlike many donor supported interventions in Africa, this initiative is also market driven with the ability to provide efficient and effective solutions to agribusinesses and policy makers.

Chapter 9 explores how ICTs have been harnessed in Local Governments in Nigeria and the possibilities the Push-ICT theory perspective offers in harnessing ICT in local government administration using Nigeria as a springboard to other African countries. It is pointed out that several international initiatives have shown that effective utilization of ICT offers new possibilities for improved efficiency in governance, new ways of citizens' engagement and a more active participation in policy-making, resulting in improving and rebuilding of trust, transformation of relations between governments and their citizens. However, there are gaps that need to be addressed to be able to explore the full potentials of ICTs. Some of challenges listed are: inadequate or lack of training and capital, limited understanding of the potential of technology and paper documents required for administrative approvals. The Push-ICT Theory implies that you push the initiative until it's accepted. The theory is a product of several years of observation of the situation surrounding the use of some information and communication technologies (such as mobile phone, computer and computer literacy training). For example the reluctance by local government staff can be addressed by policy push. Another angle of the theory is the push from ICT users (Users Push). In Nigeria the high tariff hitherto on mobile telecommunication services has compelled subscribers to push for reduction on tariffs. It is concluded that the Push-ICT Theory approach would help make ICT available and indirectly force members of the community to use them.

Chapter 10 explores how mobile phones can be used for environmental protection with a case from the African Great Lake area involving 12 countries. A short description of the African Great Lakes and their environmental and economic importance is presented with an overview of the Equatorial Africa Deposition Network (EADN), an ambitious transboundary project that tries to identify sources of eutrophication[1], a major threat to the African Great Lakes integrity. The mobile environmental framework around the EADN project is described as an example of the key role mobile phones can play as development boosters. As part of the project illiteracy has been addressed. It is shown that mobile phones are more a social than a personal asset in Africa. When using them, social interaction and cooperation are expected. Such community effort might be a crucial factor to help illiterate people access mobile phones services. Voice Response (IVR) systems can represent another way to efficiently deal with illiteracy. Those kind of spoken dialog systems have been successfully used in rural settings to allow communities with high illiteracy rates to access meaningful information to their everyday life like agricultural information. Overall the project has demonstrated that mobile phones and their impact in Africa represent a huge opportunity for environmental protection. Lack of data has been one of the major challenges science faced when trying to propose solutions to the emerging problems of the African Great Lakes. The Equatorial Africa Deposition Network is the first project aimed at identifying long range sources of nutrients and their transport patterns. Using mobile phones as a communication channel and also as a way to trace human activity will enrich the data generated by the EADN and provide better evidence to policy.

*Chapter 11*analyzes the situations with reliable energy currently being a key sustainability focus of ICT development in the Sub – Saharan region. It has been demonstrated that energy supply has some level of impact on social and economic development in developing countries particularly in many rural and semi- urban towns. Typical mobile telecom architectures in developing countries and their energy requirements are analyzed in some details. It is concluded that solar photovoltaic technology is one of the means of providing reliable power with a CO_2 reducing solution in developing countries with

[1]Eutrophication is defined as "the over enrichment of receiving waters with mineral nutrients. The results are excessive production of autotrophs, especially algae and cyanobacteria". (Correll 1998)

sufficient 'sunshine hours'. The use of solar photovoltaic technology as an alternative energy makes it possible for areas without grid power supply to enjoy the benefits of mobile communications networks. Based on detailed simulations using Ghanaian data it is illustrated that it is also a cost effective solution. Further, the need for reliable power will continue to rise as the demand for ICT innovation keeps evolving in developing countries. The issue of renewable energy is, however, not limited only to developing countries. In telecom and ICT industry, solar photovoltaic technology has a greater advantage and has enjoyed a better rate of application to date as compared with other renewable energy technology. However, achieving the optimal configuration for large scale application is still a challenge.

Chapter 12 presents that despite huge problems in the transportation sectors and potentials in applying ICT solutions most developing countries have not yet developed such ICT solutions for their traditional transportation sectors. The focus of the chapter is to analyze attempts made in Africa to develop Intelligent Transportation Systems (ITS) illustrating the situation before the adoption of ITS technologies, and what the current state of ITS development in Africa is today. It is stated that generally, there is the need to establish regional cooperation and collaboration between African countries to enable them to mutually improve upon their transportation sectors. For this to happen, there must be no traditional, cultural, institutional and political influences on any ICT/ITS development plans and projects for the transportation sectors in any country in Africa. It is, however, concluded based case studies of the situation in Ghana that the solution to developing and improving transportation sectors in Africa, is not insurmountable if efforts are put into human, technical, capital and technological resources.

Some of the general conclusions and results that appear from the book are the extraordinary growth in mobile in Sub-Saharan Africa have delivered great growth potentials and that these are not fully exploited; despite the growth there is still need for affordable and widespread access including rural areas; the results obtained some areas/ sectors are highly needed in the others with education, governance and energy as examples – this is, however, also not efficiently exploited.

List of Figures

List of Tables

1

Some Policy Considerations in the Light of Mobile Broadband Development in Africa

Alison Gillwald
Research ICT Africa & Graduate School of Business,
University of Cape Town, Management of Infrastructure
Reform and Regulation Programme
agillwald@researchictafrica.net

Enrico Calandro
Research ICT Africa & Graduate School of Business,
University of Cape Town, Management of Infrastructure
Reform and Regulation Programme
ecalandro@researchictafrica.net

1.1 Introduction

Universal access to communication services was a developing country policy issue a decade ago. Now it is a global issue - this time to high speed, affordable bandwidth. With such networks and services regarded as a necessary condition for the development of information societies and knowledge economies, the price and quality of service of broadband networks is becoming an increasingly significant policy issue.

Consumer access to high quality broadband services is predicated on national networks capable of supporting the rapid growth in traffic at competitive prices. Under such conditions research suggests that an increase of 10% in broadband penetration can produce a 1% increase in GDP (Kim, Kelly, & Raja, 2010), arguably more in developing countries.

But while universal access to broadband is a policy issue common to countries across the globe, broadband has evolved very differently in the South from more mature markets in the North, presenting quite different public policy and regulatory challenges. Assumptions about the primary means of broadband access, the cost and quality of fixed versus mobile broadband services underpinning 'best practice' broadband policy are challenged, particularly in the African context. Empirical research shows that all over African countries, mobile broadband has overtaken fixed broadband in terms of subscribers, prices and speed of service.

Definitions of broadband are increasing moving beyond the traditional notion of a specific network at minimum transmission speed (International Telecommunications Union, 2010). Rather, the World Bank suggests viewing broadband as an ecosystem which includes its networks, the services that the networks carry, the applications delivered and users. Each of these components has been transformed by technological, business and market developments (Kim, Kelly, & Raja, 2010).

Approaching broadband from an ecosystem point of view allows clearly identify differences between the broadband ecosystem in the North which is based on fixed, cable and fiber optic networks, cloud computing applications, fast and reliable cheap connectivity, and the broadband ecosystem in the South, which can be defined as a mobile broadband ecosystem made out of wireless infrastructures, mobile apps, high prices and poor quality of services.

This chapter is based on household and individual ICT access and use data, collected by Research ICT Africa between 2011 and 2012[1] in 12 African countries. In addition, it uses broadband prices collected in November 2013 across 12 African countries[2] and it draws on the two studies on measuring broadband performance in South Africa[3].

1.2 More Wireless Broadband Connections than Fixed

Unlike in more mature markets, where fixed services are the predominant broadband platform, in Africa mobile networks provide the primary means of broadband access (Stork, Calandro, & Gillwald, 2013). As with voice

[1]RIA household and individual ICT access and use surveys delivers nationally representative data in 12 African countries (Research ICT Africa, 2012). RIA ICT survey data is publicly available. Please contact info@researchictafrica.net to enquire.

[2]Broadband prices (fixed and mobile) are available at www.researchictafrica.net/broadband

[3]Two piloting studies were conducted in South Africa to assess broadband performance. The first one compared mobile and fixed speeds and latency values in South Africa (Chetty, Sundaresan, et al., 2013), while the second tracks ISPs interconnectivity (Gupta, Calder, et al., 2013).

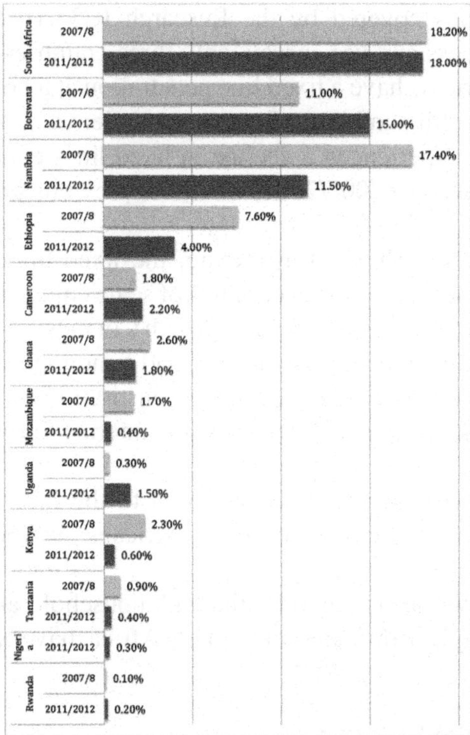

Figure 1.1 Household with working fixed-line

(Source: RIA household and individual ICT surveys 2007/8; 2011/12)

services where massive pent demand was met by through the wireless revolution that transformed communications on the African continent, demand for Internet by those unable to access the limited ADSL services available on the continent, has been met through mobile service providers. Differently from the established telecommunications markets in the North where access and where ADSL-upgradable copper networks or upgradable cable networks were almost universal at the advent of broadband, in Africa most fixed line networks reached less than 1% of the population and fiber remains negligible.

This dependence on usually, the monopoly provision of fixed infrastructure for broadband delivery constrained Internet penetration. The most wired country in sub-Saharan Africa, South Africa, following privatization and liberalisation has a declining fixed line teledensity of less than 20%.

Among the countries covered by the Research ICT Africa household and individual ICT access and use survey, only three countries – Botswana, Namibia and South Africa – have a fixed line penetration at a household level higher than 10%, while Mozambique, Tanzania and Kenya do not reach 1% penetration. They also experienced a decline in fixed lines connectivity at a household level compared to 2007/2008, when the previous household ICT survey was carried out.

With such limited fixed-line infrastructure, the fixed-line operators on the continent do not benefit from the economies of scale enjoyed by operators in the North (or by the mobile operators in the South). In most countries on the continent, historically, incumbents also usually had a monopoly on undersea cable access, with international bandwidth costs constituting up to 80% of the costs of local internet service providers (interview internet Service Providers Association, RIA 2008).

The combination of these factors contributed to the high cost and the unaffordability of internet services for most Africans, inhibiting internet diffusion.

Among African countries covered by the RIA household and individual ICT survey, only Kenya, Botswana and South Africa have an individual

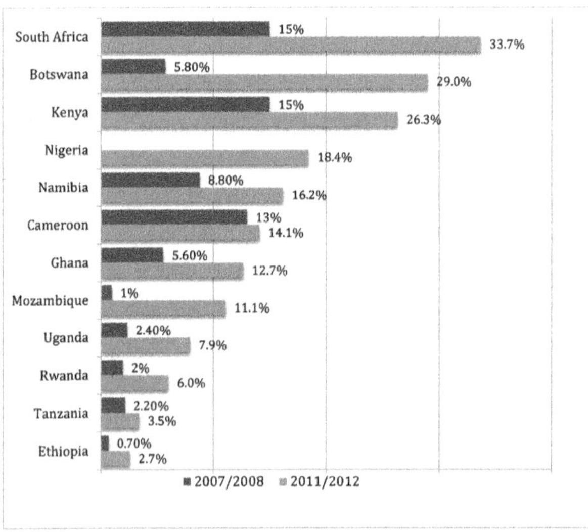

Figure 1.2 Internet use in selected african countries

(Source: RIA household and individual ICT surveys 2007/8; 2011/12) *Nigeria 2007/08 data omitted

internet use rate higher than 20% of the population 15 years old or older. In countries such as Uganda, Rwanda, Tanzania and Ethiopia, internet use is negligible, since less than 8% of the population has ever used the Internet. However, although internet use results very low, compared to the previous RIA ICT survey conducted in 2007/2008, internet use increased significantly, almost four times in Ethiopia and as much as eleven times in Mozambique.

The breaking of the SAT3 stranglehold on international bandwidth on the continent with the landing of SEACOM undersea cable in Africa in 2009, then EASSy 2010, connecting the East coast of Africa for the first time, and subsequently the terra-bandwidth from West African Cable System (WACS) and AIM in 2013, rapidly changed the pricing structure of the industry. Wholesale international bandwidth is now priced at a fraction of what it was then, even although these benefits have not always been passed on fully to end users to stimulate take up.

This both enabled and was driven by the introduction of broadband mobile technologies. Constantly reducing prices for smarter devices and service, marketing and pricing innovation fuelled the uptake of broadband services.

Considering the lower set up costs of mobile data compared to fix, particularly for low data use and uneven consumption, with no monthly line rental charges and installation fees and convenient prepaid charging options,

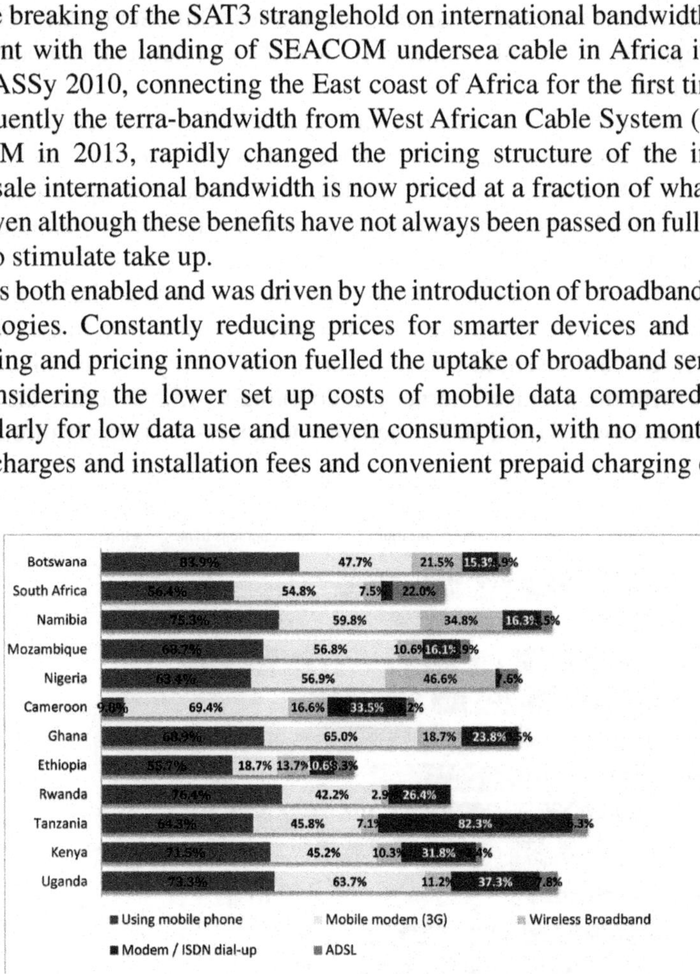

Figure 1.3 Type of internet connection by households with internet access (multiple responses).

Source: RIA household and individual ICT surveys 2007/8; 2011/12).

the dominance of mobile is unsurprising. For these reasons, many African home users opt to use a 3G dongle modem or mobile handset to access the internet instead of setting up an ADSL connection, which is often simply not available and if it is, unaffordable.

Figure 1.3 above depicts that in all countries surveyed, mobile connectivity overcame ADSL and fixed-line internet connections, which are present in less than 8% of the households surveyed in RIA countries. Only South Africa has a relatively high level of ADSL connectivity with 22% of households reached by this technology.

Although the data service vacuum has been filled by mobile operators offering 2.5, 3G and now also 4G services in some countries such as South Africa and Namibia, significant mobile broadband take up remained constrained initially by the requirement of a computer into which 2.5G and 3G dongles had to be plugged.

But the rise of feature and smart phones over the last few years has enabled growing numbers of Africans to access the internet for the first time. In the 2012 Research ICT Africa household and individual user survey we see internet going mobile with the average internet use doubling to 15.5% across the 12 African countries covered from 2008 when internet access was largely through

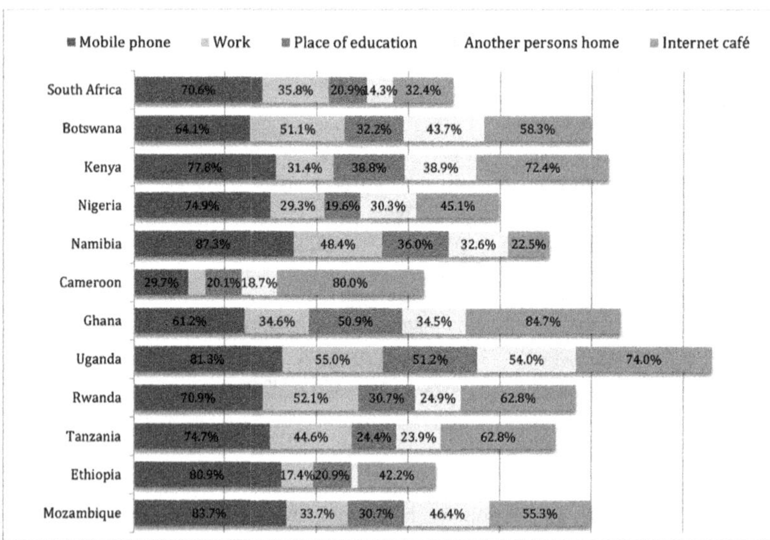

Figure 1.4 Share of where internet was used in the last 12 months (multiple responses)

(Source: RIA household and individual ICT surveys 2007/8; 2011/12).

internet cafés, schools or workplace. (Stork, Calandro, & Gillwald, 2013). As can be seen from the Table 1.1 below this is largely as a result of mobile phone access.

The mobile has become the main communications tool to access to the internet in all countries surveyed except Ghana and Cameroon. Public access facilities are still relevant access points for those who cannot afford a computer and a fixed or 3G dongle connectivity and where phone penetration remains low. In Tanzania, Uganda and Kenya more than 60% of internet users accessed to the internet at a public access facility such as internet cafés.

Not only mobile penetration has reached the critical mass in voice network, which is believed to trigger the network effects associated with economic growth but also the number of devices capable of browsing the internet has increased significantly across all these countries. However, although numbers of internet-enabled devices are growing across all these African countries, at the moment only a small portion of those devices has been used to browse the internet. The main reason for the slow uptake of mobile internet is the high cost of broadband data, which is still prohibitive especially for the poor (Calandro, Gillwald, Deen-Swarray, Stork, & Esselaar, 2012).

Table 1.1 Share of where internet was used in the last 12 months (multiple responses)

	15+ year olds that own a mobile phone	15+ year olds owning a mobile phone				
		Prepaid	Mobile is capable of browsing the internet	Use social networking (Facebook, etc.)	Browse the internet on mobile	Reading and writing emails on mobile
South Africa	84.2%	87.5%	51.0%	25.4%	27.6%	16.7%
Botswana	80.0%	97.4%	29.5%	18.4%	22.8%	16.5%
Kenya	74.0%	99.8%	32.3%	24.5%	25.3%	19.7%
Nigeria	66.4%	99.0%	22.7%	15.8%	16.0%	14.6%
Ghana	59.5%	97.4%	28.5%	11.3%	13.4%	9.5%
Namibia	56.1%	91.8%	30.7%	17.3%	23.8%	12.4%
Uganda	46.7%	98.0%	14.9%	6.7%	7.7%	6.0%
Cameroon	44.5%	99.0%	14.9%	7.7%	8.1%	4.3%
Tanzania	35.8%	99.5%	19.2%	4.7%	5.2%	5.2%
Rwanda	24.4%	90.1%	19.1%	13.6%	14.9%	13.3%
Ethiopia	18.3%	98.4%	6.5%	2.1%	5.1%	9.7%

1.3 Mobile Broadband Generally Cheaper than Fixed

Wireless mobile broadband is now the primary means of access to the internet in Africa for individuals unlike in more mature markets. The main reason for that is its lower cost compared to similar fixed-line offerings. From analyses of prepaid and contract mobile and ADSL (fixed) broadband price baskets undertaken by Research ICT Africa[4] it is clear that fixed-line packages are far more expensive than comparable mobile packages. The tables below compare ADSL offerings and postpaid and prepaid mobile broadband tariffs across 12 African countries.

Table 1.2 below shows that in the up to 1GB basket delivered as mobile broadband at a speed up to 7.2 Mbps, prepaid is generally cheaper than post-paid except in South Africa[5], Namibia and Botswana. In all these countries, however, ADSL offerings at similar speeds are up to twenty times more expensive[6] than similar mobile broadband baskets.

Table 1.2 Comparison of 1GB capacity delivered as mobile (up to 7.2 Mbps) and as ADSL (up to 4Mbps) in USD, November 2013

Country	Prepaid	Post-paid	ADSL
Tanzania	3.20	10.40	64.00
Ghana	9.26	11.98	
Kenya	10.01	11.02	122.18
Uganda	11.14	16.29	
Rwanda	14.19		
South Africa	14.61	9.63	53.05
Mozambique	18.80	22.38	155.14
Nigeria	20.10	20.10	
Cameroon	20.50		397.36
Namibia	33.52	30.25	121.87
Botswana	81.21	26.70	
Ethiopia		17.61	

Note: Usually mobile broadband speeds in the 7.2Mbps basket are offered at a speed of 3.1 Mbps (EVDO).

[4]See http://www.researchictafrica.net/prices/broadband.php

[5]The lower cost of contracts than prepaid plans in South Africa is due to the relatively high number of postpaid customers compared to other African countries. Competitive forces in the postpaid market segment have pushed mobile operators to reduce postpaid broadband prices. See RIA Policy Brief SA No. 2, 2013 (Research ICT Africa, 2013b).

[6]Although ADSL packages differ from mobile broadband packages in terms of speed offered and billing system, in order to compare them we selected mobile broadband and fixed packages which have similar speeds.

Table 1.3 Mobile access and use in selected african countries

Country	Prepaid	Post-paid	ADSL
Tanzania	6.80	29.60	64.00
South Africa	20.32	26.35	53.05
Cameroon	20.50		397.36
Ghana	22.84	33.97	
Rwanda	25.88		
Kenya	34.05	36.80	122.18
Uganda	40.59	41.60	
Nigeria	48.66	51.84	
Mozambique	54.60	54.60	155.14
Namibia	119.07	43.67	121.87
Botswana	401.64	308.41	
Ethiopia		65.96	

Note: Usually mobile broadband speeds in the 7.2 Mbps basket are offered at a speed of 3.1 Mbps (EVDO).

ADSL is much more expensive than mobile in all the countries tracked, except Botswana, where 1GB prepaid baskets at a speed up to 7.2 Mbps is more expensive than similar postpaid ADSL offerings. Looking at 1GB only in Table 1.3, it is evident that the mobile operators with the most competitive pre-paid voice services on the continent also provide the cheapest pre-paid data (RIA, 2013a).

Table 1.2 above illustrate that in the up to 1GB basket delivered as mobile broadband at a speed up to 7.2 Mbps, only a few countries match their prepaid and postpaid offerings, although postpaid results generally more expensive. In Tanzania, postpaid is three times more expensive than prepaid. By contrast, in Botswana postpaid is three times cheaper than prepaid and in South Africa the cheapest prepaid basket is 1.5 times the cost of a similar postpaid basket. Mobile broadband prepaid prices in Mozambique, Nigeria, Cameroon, Namibia, and Botswana are inhibiting broadband take up, while in Botswana pre-paid prices are simply extortionate. South Africa, the only country with significant numbers of users in the more lucrative post-paid (contract) market sees significant competition for these high-end customers who enjoy some of the best prices. As a result South Africa offers the lowest contract prices for 1, 5 and 10 GB.

With mobile operators offering different broadband data options and packages for different market segments, the mobile broadband market is where pricing pressure is occurring, as operators compete to attract and retain broadband customers. As a result where fixed operators have moved into the mobile market with the intention of benefiting from bundled service offerings the danger exist of their mobile services, which may be cheaper and of better quality as a result of often fierce competition in the mobile market, cannibalizing their fixed line offering.

This is evident in South Africa, which, although limited has the most extensive fixed network in sub-Saharan Africa. In that country mobile operators, particularly its late entrant mobile arm, Telkom Mobile, offer cheaper and faster internet plans than the data service offered by the fixed incumbent, Telkom.

1.4 Mobile is faster than Fixed Broadband

Unlike in mature markets, mobile broadband is not only cheaper but faster than fixed in the growing number of markets deploying third and fourth generation mobile wireless technologies. This was revealed in two studies conducted between February 2013 and July 2013 in South Africa to assess broadband performance. The first study (Chetty, Sundaresan, Muckaden, Feamster, & Calandro, 2013) provided a systematic assessment of both mobile and fixed-line broadband performance in South Africa. On the one hand it investigated whether the speed of fixed and mobile broadband services achieves the performance advertised by the ISPs in South Africa; on the other, it compared fixed broadband performance to mobile. The study also revealed the latency that users in South Africa experience to commonly accessed web sites.

These studies based on measuring broadband performance through router and mobile apps speed testing have shown that mobile speeds are faster than fixed-line speeds, irrespective of whether they are connecting to a local or international server, when comparing similar packages. The median downstream throughputs of the LTE connections are the highest, at around 25 Mbps. The 3G connections also generally experience higher download throughput than the fixed-line service plans. These are confirmed using host-based measurements from MyBroadband reflected in the Figure 1.5 below. The download speed of mobile broadband is far outperforming fixed-line broadband speed in South Africa. Only with regard to latency does fixed marginally outperform mobile. High latency deteriorates user experience of playing 3D games, watching high quality videos, but also VoIP calls, and it might be an obstacle to the uptake of real time services requiring stable connectivity such as cloud computing.

This research put some light on the fact that throughput is not the only broadband performance problem. Equally important is the latency to websites

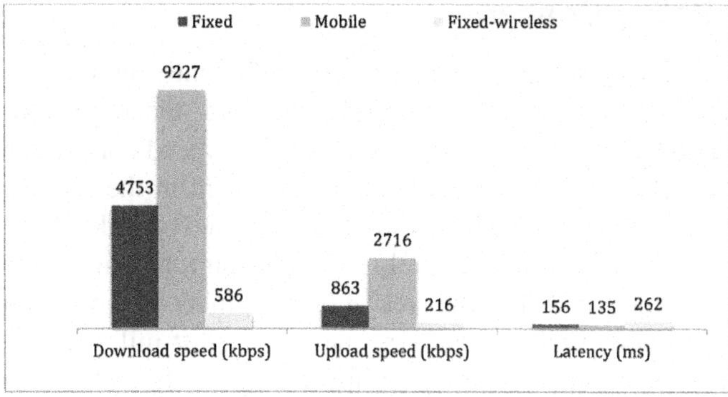

Figure 1.5 Fixed and mobile throughput and latency.

(Source: RIA presentation of MyBroadband data: 75,000 measurements in 2013 across South Africa.)

and services that users actually visit. The research suggested that both the physical distance to servers for popular sites and the ISP-level of connectivity between users and these servers could introduce latencies of several hundred milliseconds (Chetty et al., 2013). From the first study it became clear that not only speed resulted particularly poor, but also latency deteriorates broadband performance.

Based on the results of this first study, the second study (Gupta, et al., 2013) investigated the causes of circuitous internet paths and evaluated the benefits of increased peering and better placed proxy placement for reducing latency to popular internet sites. Research results show that ISP interconnectivity in Africa is circuitous[7]. African ISPs, although they might have a presence at African IXPs, do not peer at these IXPs. Rather they hub in Europe because of the economy of scale[8] before reaching final destinations in Africa. Further, the absence of intra-African IP traffic incentives African ISPs to interconnect in the continent. As a result, the latency of these paths is significantly high (Gupta, et al., 2013).

[7]A circuitous path is one that traverses a point in the internet physical infrastructure that is far from the path created by taking the geographically shortest path between two endpoints.

[8]The main reason for African ISPs to peer in Europe at LINX, for instance, is that most of the ISPs they need to peer with are present and peer at LINX, but not at African IXPs'.

1.5 Arising Policy Concerns

With dependence on mobile wireless, which is inherently less stable than fixed broadband technologies such as xDSL and fibre, the implications of not having ubiquitous, reliable always-available, high-speed connectivity for the economy and global competitiveness are significant in the longer term. Data services (such as Skype) are increasingly being used to make voice calls and SMSing is giving way almost entirely to free instant messaging services such as WhatsApp and BBM. But the lack of development of always-on high speed and quality bandwidth in the access networks (last mile) required by business, public institutions and citizens impacts negatively on the countries informational development as a key determinant of global competitiveness.

1.5.1 International Bandwidth, National Backbone and Backhaul Networks

In the meantime the immediate relief provided by wireless and mobile services to bandwidth starved Africans, has resulted in a massive rise of data traffic. The challenges in the various parts of the broadband value chain differ somewhat in different African countries but generally the critical infrastructure issue is no longer international bandwidth, as it was at its best with a single monopoly provider, only five years ago in most parts of the continent. The challenge is now the development of the terrestrial backbone and backhaul networks. Historically dimensioned for low bandwidth voice services, the current capacity of these networks is extremely strained in many countries.

Debates on whether improved broadband penetration is best achieved by facilities-based competition or by avoiding infrastructure duplication through the consolidation or building of national open access broadband networks on which service based competition can be enabled, rage on. In many countries new fibre backbones have been built through long-term concessions to supplier builders, or build operate transfer models, or public infrastructure loans from development banks. In many countries, particularly in East Africa these are articulated to ensure access by landlocked countries to undersea cables out of Kenya.

1.5.2 Access Network

The access networks continue to be a major challenge. Although primary access to broadband is through mobile, several countries do not even have fully rolled out 3G networks. Dependent on EDGE technology their experience of broadband is limited by low speeds and high latency. The release of 4G spectrum has been delayed in many countries and many countries continue to struggle with the best way to assign it. Although operators in some countries have creatively re-farmed existing spectrum in order to offer next generation spectrum (LTE) access networks, access to this high-speed technology has also been stymied by the lack of access to optimal spectrum in many parts of the continent. The institutional challenges associated with the allocation of spectrum and the migration of analogue terrestrial broadcasting to digital have meant that service innovation, tax revenues and potential job opportunities have been squandered in the meantime.

The efficient use of spectrum to meet this unprecedented demand is vital and the cost of not doing so is high. The rough assessment of the negative economic impact of the failure to release high-demand spectrum by doing a reverse application of the formula of the World Bank study linking the extension of broadband by 10% to a 1.5% increase in GDP has been equated to a loss of hundreds of billions of dollars over a ten year period. [9]

1.5.3 Broadband Pricing and Affordability

With growth rate of data revenues outstripping traditional voice revenues and clearly becoming the major source of revenue for the future, mobile broadband is where the competitive pressure in the broadband market is focused. But while African consumers with access to broadband services are benefitting from stiff price competition between mobile operators seeking to attract and retain data customers, it should not be assumed that the uptake of mobile broadband services and the price competitiveness of these in this more liberalized segment of the market in most African countries, means that issues of affordable access are addressed.

The price of broadband in Africa remains a serious challenge, not only because of its negative impact on the exponential growth in broadband required for Africa to catch up to the rest of the world. It is also because of the high input cost it represents to enterprise on the continent and the

[9] Based on recent studies linking the extension of broadband to economic growth, the GMSA concludes in its submission to the public hearings on spectrum that the ICASA proposals for assigning the new spectrum would lower the net present value of GDP by between ZAR 450 and ZAR 510 billion over the period 2014 to 2025, when compared with assignment of the spectrum to existing operators and would reduce the net present value of government tax revenues by between ZAR 95 billion and ZAR 110 billion and is equivalent to 500,000 additional jobs, at current South African wages (GSMA submission to ICASA 2012, www.ellipses.co.za).

negative consequences for growth, development and global competitiveness. The variably poor broadband penetration levels across countries are primarily an outcome of high prices. The high prices and poor quality of ADSL services reflect the absence of competition in this segment of the broadband market, in contrast to the more competitive mobile segment of the broadband market. The cost drivers underlying high fixed-broadband prices need to be identified. While international bandwidth prices, once the major factor in South African data prices, have plummeted, terrestrial and IP transit prices are now major cost drivers. The impact of these factors on the cost of communications requires regulatory assessment.

1.5.4 Quality of Service

In Africa no systematic assessment of fixed and mobile broadband performance has been conducted. National regulatory authorities by and large do not monitor quality of service and if they do so they do not regularly report on their findings. Measuring broadband performance is further complicated by the fact the mobile phone is increasingly becoming the key entry point for internet adoption on the continent and therefore monitoring mobile broadband is as much as relevant as monitoring fixed broadband performance. Further, quality of service studies on broadband performance in South Africa suggest that measuring broadband quality of service allows for the identification of factors at an infrastructure level which induce significant performance bottlenecks. On this note, the ISOC (2013) recommends the continuous monitoring of RTT and traceroutes to popular destinations, in order for the ISPs to rapidly take action either to fix networks or to contact a third-party content provider to address high latency issues.

1.5.5 Public Private Interplay and Regulatory Interface

The scale of investment required to build-out next generation networks, means that particularly in developing economies neither the state nor the private sector on their own are able to meet the broadband needs of countries in increasingly information dependent economies. This reality has resulted in new interplay between the state and market in several jurisdictions creating new delivery, investment and business models.

Early evidence from some of these national experiments emerging from the different models is that partnerships between the state and private investors and operators can be key enablers of broadband infrastructure development. As in the past, success remains dependent on high levels of state co-ordination, sophisticated skills set within government and more flexible bureaucratic processes. Those are neccessary conditions to manage new innovative funding models, together with well resourced, capable national regulatory agencies, with the correct institutional arrangements to enable this.

1.6 Conclusion

While universal access to broadband has become a policy issue common to countries across the globe, broadband has evolved very differently in the South compared to more mature markets in the North, presenting quite different public policy and regulatory challenges. Assumptions about the primary means of broadband access as well as the cost and quality of fixed versus mobile broadband services underpinning 'best practice' broadband policy are challenged, particularly in the African context. Major policy challenges remain for Africa if it has to create the conditions for investment in backbone and backhaul networks to deal with the demand for high-speed spectrum and fibre access networks required for an increasing member of users. Achieving an enabling policy and regulatory framework conducive to investment, the rationalisation of existing infrastructure and the coordination of infrastructure planning going forward are a key challenges for African countries wishing to develop their societies and economies and become globally competitive.

2

The Impact of Liberalization on the Mobile Telephony Market in Africa: the Cases of Ghana, Nigeria and Kenya

Idongesit Williams
CMI, Department of Electronic System
Aalborg University, Copenhagen, Denmark
idong@es.aau.dk

Benjamin Kwofie
CMI, Department of Electronic System
AalborgUniversity,Copenhagen, Denmark
bkwo@cmi.aau.dk

2.1 Introduction

The development of mobile telephony has come a long way from the 1980s till date. It has been noted that the mobile network coverage in urban Africa rose from 16% in the early 1990's to 90% in ten years (Forster & Briceno-Garmendia, 2011). From the same source, within the same time period, Mobile telephone networks coverage exists in fewer than 50% of rural Africa. From the colonial days, telecommunication technology in Africa (in general) has evolved from the telegraph to the circuit switched Plain Old Telephone Service (POTS) to digital telephony. Today, Mobile broadband enabled by 3G and 4G Mobile standards exist in the African Market. Currently, the move is towards Long Term Evolution (LTE). LTE currently exists in some African countries like South Africa, Rwanda, Angola and Tanzania (All Africa, 2013).

The growth of mobile telephony in Africa isn't accidental. The demand for some form of remote communication services was there, especially with the inefficiency of some Postal Services in sub-Saharan Africa. African families are closely knit units, although today these units are scattered, they still communicate and support each other. Hence the need for an efficient remote communication was there. Unfortunately, the supply of fixed-line and mobile telephony was very low. Low quality of service, delayed maintenance when the telephone had problems, high tariffs were some problems that inhibited African users from adopting telephony. Therefore telephony wasn't the best alternative to the postal service. Analogue mobile-telephony had more spectrum requirements and it was not possible to extend coverage to larger areas at a cheaper cost. However once digital mobile standards (2G standards and above) was introduced into the African markets, the supply for telephony (mobile telephony) improved greatly. Digital mobile telephony required less bandwidth and the service could be extended to a greater area. Digital mobile telephony infrastructure was cheaper to deploy and extend. There were possibilities of more telephony services besides voice service. There was also a possibility of delivering the mobile telephony services via innovative business models. Hence the advancement in mobile technology led to the availability of mobile telephony and the affordability of the service. With these advancements, users could meet their communication needs.

The entrance of digital mobile telephony (2G and above mobile standards) in Africa is also not accidental. The market reforms introduced to Africa by the international development partners in the late 1980s and 1990s was the catalyst to entrance of digital mobile telephony (Wallsten, 2001). The idea of the market reform was to create a paradigm shift in public infrastructure development. Initially, public infrastructure development was a prerogative of the public sector. But with market reforms, the public's role in infrastructure development was shifted from government to governance. Some market reform initiative in Africa included privatization and market liberalization. Mobile telephony infrastructure development has improved greatly in Africa as a result of the telecom market liberalization. With liberalization, there was the removal of market entry barriers to Mobile Telephone Operators in sub-Saharan African markets.

This chapter tells the story of the mobile market liberalization in Africa. It starts with a general overview of mobile market liberalization in Africa and later converges towards three specific cases. Issues discussed include the mobile telephony penetration before and after telecom liberalization in Africa. In discussing the general and specific issues, the drivers that led to the growth of mobile telephony are mentioned. The specific telecom market liberalization cases chosen are Ghana, Nigeria and Kenya. The reason for the cases are as follows:

- Nigeria is the fastest growing mobile market in Africa.
- Ghana is one of the first countries to introduce mobile telephony in sub-Saharan Africa and they also have a high mobile telephony penetration rate.
- Kenya is one of the countries in sub-Saharan Africa whose privatization effort, although it started on a shaky note, could transform the incumbent into 2 competing companies with innovative mobile solutions. Kenya's difference from the rest also lies in the ability to incorporate PPP in its regulatory effort.

2.2 Market Reforms

2.2.1 Short Overview of Market Reforms in Africa

Telephony development under the Post and Telecom national monopoly regime was bridled with inefficiency. The inefficiency was as a result of the following:

- **Regulatory capture**: This existed with regards to the regulation of national monopolies (Melody W. H., 1997). Here the national monopoly had no independence as a result of the fact that they were national assets. Hence the management of the network operators were either civil servants or political appointees.
- **Political interference**: Bases on the previous point the political appointees were not able to make professional choices as they were hampered by political interference in their decision making (Haggarty, Shirley, & Wallsten, 2002).

The inability to make professional decisions led to the poor and inefficient development of fixed-line telecom infrastructure and inevitably poor service delivery (Wallsten, 2001) (Sumbwanyambe & Nel, 2011). The poor service delivery was based on the inability of the incumbent operators to efficiently maintain, manage, operate and expand their networks was internal management conflicts and poor access to capital (Frempong G., 2002).

To curb these inefficiencies market reforms were introduced. These market reforms in Africa were legislative which resulted, in part, in institutional reforms. These reforms were mainly aimed at attracting public investment for funding public infrastructure. The most prominent reforms in the telecom industry in Africa were initially, commercialization and later Privatization and Liberalization.

The market reforms resulted in the following:

- The Split of P&Ts: The P&Ts were national monopolies, operating as an arm of government, were split into two. A national regulator was also established to regulate either the telecom sector alone or a combination of other sectors as well. Initially P&Ts were either self-regulatory or regulated by a ministry, for example, ministry of communications or ministry of technology etc. depending on where the government of the day decided to place it. These were important national assets.
- Policy initiatives: The legislative reforms in many cases led to the subsequent creation of Information Communication Technology for Development (ICT4D) policies, telecommunication policies, broadcasting policies, Electronic communication policies and in recent times, Broadband policies among others. The early ICT4D policies were wish lists and strategies on how to achieve the content of this lists. For most telecom policies, the emphasis was in the efficient delivery of fixed-line telephony. Mobile telephony and Broadband were mentioned in this policy but not much emphasis was given to it. At that time the penetration of fixed-line telephony was greater than that of mobile telephony. Public Institutions in Africa did not foresee the rapid growth of mobile telephony and broadband at that time. These policies introduced initiatives such as privatization, liberalization or deregulation, and in some cases commercialization.

2.2.2 Privatization

Based on this fact, national telecom monopolies in Africa underwent mostly partial-privatization. The idea behind privatization for the public sector was; the private sector will be more efficient in managing the infrastructure than the public sector. For the private sector, some determinants for participating in the privatization were: wealth, population distribution, geographical location, political accountability, country risk and telecom status in the country (Gasmi, Maingard, Noumba, & Recuero Virto, 2011). However, population distribution and telecom status were greater determinants in private sector participation in Africa than the other determinants. Most initial privatization efforts in Africa failed as a result of country risk. Some identified country risk include:

- There have been accusations of lack of transparency, an examples is the initial privatization of Telkom Kenya (AfriCOG, 2010).

- Undue political interference. An example is the Nigerian Pentascope deal, (Odemwingie, 2013).
- Financial risk due to huge debts owed by the incumbent monopoly and the incumbent monopoly's dilapidated infrastructure. Hence foreign investors in some cases were not so willing to invest as much as they promised in the development of these network infrastructure due to the financial risk involved (Alhassan, 2007).

Although the NITEL deal did collapse, the Telkom Kenya deal survived. However in recent times, the Telkom Kenya deal is still being questioned by the Kenyan press (Kaara, 2013).

It is really difficult to say that the privatization of the state monopolies was or is really successful. Success here has to be defined by the user. But one can't overlook the fact, that although insignificant, privatized national monopolies have contributed to the growth of mobile telephony in Africa.

Privatization so far in Africa has been upheld on trial and error basis. Many African countries like Ghana, Uganda and Botswana etc., have had to embark on trial by error privatization initiatives till they could get one stable investor. Some of the Privatized Sub-Saharan companies have also made effort to buy stakes in other national incumbent networks. An example is Telekom SA's bid for Uganda telecoms (Muwonge & Gomes, 2007).

2.2.3 Liberalization

The Market reform that had a great impact in Africa, enabled by the policies mentioned earlier was market liberalization. But market liberalization on its own did not have an impact until the advent of the digital mobile standard (2G and above standards) as mentioned earlier. The Global System for Mobile communications (GSM) became dominant because the manufacturers and suppliers of European brands of GSM equipment were already dealing with the incumbent operators. They had their sales outlet in Africa and to ensure interoperability the incoming network operators had to deal with these existing telecom equipment suppliers and manufacturers. Hence there was an interplay between technology, policy and market. It is believed in academic circles that the development of the telecom sector these days is affected by this three factors (Melody W., 2009) (Williams, 2012).

Liberalization of the mobile markets in Africa changed the landscape of mobile telephony in sub-Saharan Africa. However telecom liberalization in Africa didn't take off as soon as the sector reform legislations were made as seen in the table below.

Table 2.1 Adoption of market reforms in africa

Country	Year reform legislation passed	Year Independent regulator established	Year Incumbent privatized	Number of mobile competitors by 1997	Number of mobile competitors by 2013
Botswana	1996	1997	2012	0	4
Cameroon	1998	2010	–	0	4
Cote d'Ivoire	1994	1996	1997	3	5
Ghana	1994	1997	1997	3	6
Kenya	1994	1998	1999	0	4
Malawi	1998	1998	2005	0	2
Morocco	1996	1997	2001	0	3
Mozambique	–	1992	–	0	3
Nigeria*	1992	1992	2001	0	9*
Senegal	1996	–	–	0	3
Tanzania	1994	1994	2001	1	11
Uganda	1994	1997	2010	0	8
Zambia	1994	1994	2010	1	3
South Africa	1993	1997	1997	2	4

Sources: 1. (Wallsten, 2001) Extracted from ITU (1998a, b, c), Pyramid Research (1997), U.S. Department of Commerce (1996), 2. (Budde a, 2013) (Budde b, 2013) (Budde c, 2013), (BOCRA, 2013) (BTC, 2013), (TRB, 2013), (ART, 2013) (LusakaTimes, 2010), (UCC, 2013)
*There are more companies but with subscriber rate below 100 000 subscribers

The reason for the delay was because of the late setting up of telecom regulators. Although administrative procedures were used to grant licenses to incumbent private mobile operators, the establishment of the regulatory authorities provided an independent platform for telecom regulation. The liberalization of mobile market in Africa has led to competition in the provision of mobile telephony services. The competitive market led to innovation in regulations and in some cased regulatory forbearance. As the companies competed among themselves, new business models emerged which led to tariff reduction. Tariff reduction helped in the growth of demand for mobile telephony in Africa but it is difficult to say that it helped the supply that much.

Although liberalization was aimed at the removal of regulations, as competition emerged in the African market, there was need in some cases to regulate competition to ensure protect the overall public interest of universal Access and Service. Some regulatory initiatives included:

- Market entry and licensing: The lowering of market entry requirements enabled the entrance of international bandwidth providers as well as national bandwidth providers (MNOs) into the African market. As seen in Table 2.1 above, in 1997, most sub-Saharan African countries had not more than 3 mobile network operators. By 2013, most countries could record more than 10 mobile networks operating in Africa. MTN (An African MNO operating in Africa and the Middle East) is the largest mobile operator in Nigeria. Aside Millicom and a few smaller MNO brands, Asian, Middle East, European and a few African MNOs operate in the African market.

- Gateway Liberalization: Tanzania, Botswana, Kenya, Morocco, Nigeria, and Ghana are some countries that liberalized their international gateways. With liberalized gateways, MNO's had the option of which gateway to connect with and that produced competition and led to the reduction in pricing and not very effectively but in a way reduced international interconnection problems. However some Second national operators like the one in Ghana as will be seen later didn't provide any competition to the incumbent. In Tanzania between 2001 and 2005, tariff for international calls dropped by 49% from $2.74 to $0.45 per minute (Lazauskaite, 2008). There are similar outcomes in Botswana, Nigeria etc.
- International bandwidth liberalization: In Africa today aside land locked countries, some countries along the coasts play host to a minimum of two international bandwidth providers. Until recently, Vodafone SAT3 cable owned by Vodafone served international bandwidth to most African countries. However, today GLO1, TEAMS, EASSy, SEACOMS, LION and others are providing International Bandwidth to African nations. Ghana for example is served by 3 international bandwidth providers. South Africa and Nigeria are served by 4 international bandwidth providers respectively. With the international bandwidth liberalization, the national gateways can access connectivity at a competitive price. In Ghana for example, with the Vodafone SAT3 monopoly broken, the cost of E1 bandwidth connection has reduced from $4500 to less than $1000 (GISPA, 2012).
- Introduction of converged or unified licensing: Unified licensing is another regulatory incentive that has enabled the development of mobile in Africa. In the case of unified licensing as in Nigeria, Internet Service providers, fixed wireless operators are granted license to provide telecom services in a particular area. Tanzania and Nigeria are some African Countries known for adopting converged or unified licensing in Africa. In Nigeria, the reason they have such a large number of operators in

the mobile telephony market is because of the Unified licensing regime granted to the network operators. This licensing regime enabled operators that were only fixed wireless operators like Reltel and Starcoms to provide CDMA mobile solutions to some part of the country. Hence there is room for the big five operators (MTN, Glo, Etisalat, Airtel and Visafone) with subscriber rate between 2 million and 55 million in between them. Also there is room for smaller operators providing fixed wires services. For a large market, such as Nigeria, such an initiative proved effective and ensured competition. However the flip side of it is that the smaller companies at some point get phased out by competition.

- Technology neutrality: Some African countries have adopted the technology neutrality approach to ensure Universal access and service. Aside South Africa, which is not technology neutral, Countries like Nigeria, Ghana are technology neutral. Technology neutrality has positive effect in terms of encouraging innovative Service delivery. For an example, few years after the adoption of 3G, which is still in its infancy in terms of diffusion and adoption in sub-Saharan Africa, South Africa (not technology neutral), Kenya, Ghana, Nigeria and a few others have deployed WiMAX in a small scale compared to their population. LTE has been tested in Kenya and South Africa and a few other countries, but it is in its infancy stage. The negative side of this is that cities alone will benefit from the technological innovations, leaving rural areas lagging behind in modern mobile infrastructure development.

Liberalization was also accompanied by market interventions. These interventions include, tax holidays, Public infrastructure development (back haul fiber-optic networks, last mile microwave tower for collocation etc.). In Nigeria and Ghana, telecom companies enjoy tax holidays for the 7 years and 5 years respectively if the company invests as pioneers (NIPC, 2013) (GIPC, 2013). Most African countries have established Universality funds to aid in the extension of the market to rural areas.

Liberalization of the telecommunications market in Africa has led the development of a competitive market in most African countries. The end result of the competition includes, reduction in mobile telephony tariff, as seen in Table 2.2 below, competitive business models, and innovative mobile service delivery. In the specific cases, more on the business models and services will be mentioned.

Table 2.2 Overview of mobile tariff in africa 2013

Country	Prepay Retail Price per Minute	Country	Prepay Retail Price per Minute
Kenya	$0.04	Burkina Faso	$0.16
Angola	$0.04	Cote D'ivoire	$0.17
Ghana	$0.06	Malawi	$0.17
Rwanda	$0.08	Botswana	$0.18
Nigeria	$0.08	Central African Republic	$0.19
Gambia	$0.09	Cameroon	$0.19
Uganda	$0.10	Mali	$0.21
Tanzania	$0.10	Togo	$0.23
Benin	$0.14	Chad	$0.25
Namibia	$0.14	Madagascar	$0.30
Burundi	$0.14	Gabon	$0.32
Mozambique	$0.14	Cape Verde	$0.34
Guinea Bissau	$0.15	Lesotho	$0.42
Congo	$0.16	Liberia	$1.71
South Africa	$0.16		

Source: Safaricom

In order to get a clear picture of the impact of liberalization, it is important to have a look at the pre-liberalization mobile penetration in Africa and also the post-liberalization mobile penetration in Africa.

2.2.3.1 Pre–Liberalization Level of Mobile Penetration in Africa

Initially mobile telephones were status symbols in sub-Africa. The demand for mobile telephony, provided by P&Ts, were from businessmen and corporations who had the need to communicate on the move. These mobile phones were given to employee for official purposes. Most African countries adopted market reforms in the early 1990's. Most African markets were not liberalized before 1996. Also in 1996, the growth of the fixed-line telephony is believed to have peaked in Africa (Minges, 1998). However it was in 2001 that the growth of mobile telephony outpaced the growth of fixed line telephony (ITU, 2004). Table 2.3 presents the mobile penetration levels in Africa by 1996.

Mobile vs fixed

As seen in Table 2.3 above, no African country had up to a million mobile telephone subscribers by at 1996. The number of mobile subscription in South Africa was almost close to the million subscription mark. Mobile penetration rate was relatively higher in the Republic of South Africa and in the North

Table 2.3 Mobile penetration in africa by 1996

Countries	Mobile Subscribers (1996) (Million)
South Africa	0.50000 – 1
Morocco	0.40000 – 0.49999
Mauritius	0.20000 – 0.39999
Algeria, Cote' d Ivories, DR Congo, Ghana, Nigeria	0.10000 – 0.19999
Egypt, Tunisia, Gabon, Namibia, Tanzania	0.05000 – 0.09999
Angola, Benin, Cameroon, Congo, Gambia, Kenya, Lesotho, Madagascar, Malawi, Mali, Namibia, Senegal, Seychelles, Sudan, Uganda, Zambia	0.01000 – 0.04999
Burundi	0.00500 – 0.00999
Burkina Faso, Central African Republic, Djibouti,	0.00100 – 0.00499
Equatorial Guinea	0.00050 – 0.00099
Libya, Botswana, Cape Verde, Chad, Comoros, Eritrea, Ethiopia, Guinea-Bissau, Liberia, Mauritania, Mayotte, Mozambique, Niger, Rwanda, Sao tome & Principe, Sierra Leone, Somalia, Swaziland, Togo and Zimbabwe	none

Source: (Minges, 1998)

Africans countries. The difference between the uptake of mobile in South and North Africa to sub-Saharan Africa was as a result of the following.

1. They (South Africa and North Africans) had higher Gross national incomes (World Bank. a, 2013), hence they could afford the service.
2. Almost half of their citizens lived in urban areas (World Bank .b., 2013), hence they had need and knew why they needed the service.
3. Literacy rates in these countries were higher than that of sub-Saharan Africans (World Bank c., 2013), hence they could use the service.

On the supply side the high spectrum needs, high cost of subscriber access, high monthly tariff, high cost of installation and high cost of Customer Premise Equipment (CPE) were inhibiting factors for the diffusion of mobile telephony in sub-Saharan Africa. Hence the circumstances for a supply push were more favorable for the North Africans and the Republic of South Africa than sub-Saharan Africa. It is also important to note that an plausible factor to South Africa's mobile growth as well that South African Telecom Equipment manufacturers were local (Horwitz, 1999). In North Africa, there was direct government investment in the development of public telecom infrastructure.

2.2.3.2 Post- Liberalization Level of Mobile Penetration in Africa

As the market reform legislations swept across Africa, the picture changed as well. In Figure 2.1 below, one can see the mobile subscription growth of less than 10 Million mobile subscriber lines across Africa in 1999 to almost 70 million subscribers in 2005.

The tide in mobile subscription in sub-Saharan African countries has changed today. In Nigeria, for example, the mobile subscription in Nigeria grew from about 10000 mobile subscribers in 1996 to 112 million active subscribers in 2012 (NCC, 2013). By September 2013, there were 118 million active mobile subscriptions for GSM and 2 million CDMA mobile subscriptions (NCC, 2013). Although some individual North African countries (Egypt, Algeria, and Morocco) still maintained the lead on mobile subscriptions compared to individual sub-Saharan aside Nigeria, the sub-Saharan Africans are not far behind as seen in table 2.4 below.

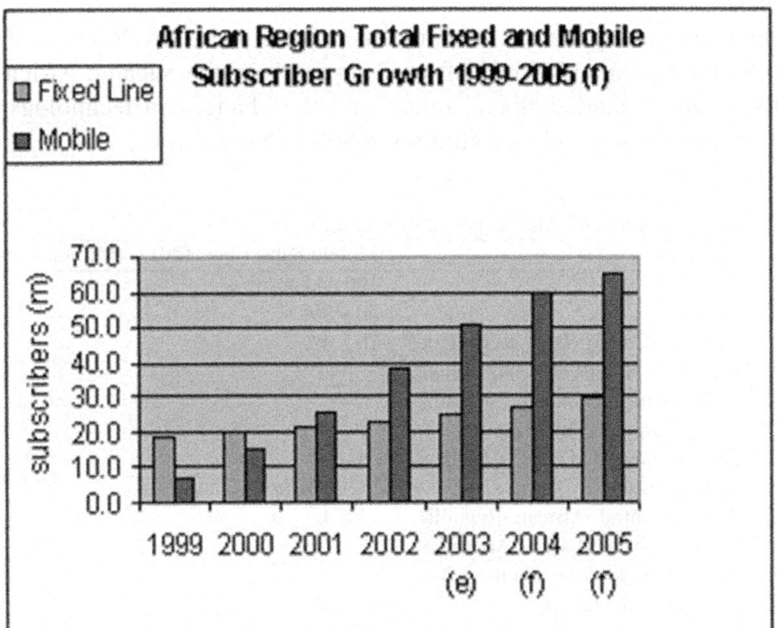

Figure 2.1 African Region Total Fixed and Mobile Subscriber Growth 1999-2005

Source: (ITU, 2004)

Some of these countries, theoretically could be said to have recorded 100% mobile penetration. Ghana is a country of 24 million people with more than 25 million mobile subscriptions according to the country's regulator (NCA, 2013). Egypt is another country that seemingly records more than 100 percent mobile subscription with 96 million mobile subscription from a population of 84 million people. Although it is difficult to ascertain if these penetration levels are a result of one man-one mobile phone, the data is still promising.

Aside the growth in mobile subscription in sub-Saharan, liberalization became a catalyst for the development of mobile in countries where no mobile network existed. In 1996 as seen in Table 2.3 above, Zimbabwe and, Ethiopia (Africa's second most populous country) had no mobile subscription. Today as seen in Table 2.4 below, Zimbabwe and Ethiopia record 12 million and 20 million mobile subscribers respectively.

The growth of mobile in Africa between 1996 and 2012 (16 years) has been phenomenal as seen in Table 2.4 below. However there is room for improvement.

Although one can say that the same factors that led to the development of mobile in South Africa and North Africa, is at work in sub-Saharan Africa today, globalization, market liberalization and the changes in technology played a greater role to mobile penetration in Sub-Saharan Africa.

Table 2.4 Mobile penetration in africa by 1996

Country	Mobile Subscribers (2012) (Million)
Nigeria	100 and above
Egypt, South Africa,	50 – 99
Algeria, Cameroun, Congo (Dem Rep), Cote d'Ivoire, Ethiopia, Gambia, Ghana, Kenya, Morocco, Mali, Senegal, Sudan, Tanzania, Tunisia, Uganda, Zambia, Zimbabwe	10 – 49
Angola, Benin, Rwanda, Madagascar, Malawi, Mozambique, Niger, Libya,	5 – 9
Botswana, Chad, Central African Republic, Congo, Guinea, Guinea Bissau, Mauritius, Mauritania, Togo, Sierra Leone, South Sudan,	1 – 4
Somalia, Equatorial Guinea,	0.5 – 0.9
Djibouti	0.1 – 0.4

Source: extracted from ITU (mobile cellular telephone subscription)

2.3 Case Studies

Case 1: Ghana

Ghana is currently a country of more than 24 million citizens. The current mobile penetration, statistically, is above 100%. However this does not imply that everyone in Ghana owns a mobile phone. One plausible reason for the statistical data could be that people own multiple SIM cards. The Ghanaian economy today has evolved from a pure mining and agricultural economy to an oil and service economy. The country's economic standing has been rebased recently from a lower income economy to a middle income economy. Before 2005, Ghana was an agricultural economy with Cocoa as the chief cash crop.

2.3.1 Pre-Regulator Mobile Market in Ghana

Ghana was the first country in sub-Saharan Africa to launch mobile telecommunications in 1992. In 1992 when the reform legislation was passed, three mobile telecom companies were awarded license to operate in Ghana via administrative procedure. These were Millicom Ghana ltd (Mobitel), Kludjeson International (CellTell) and later Scancom (Spacefon).

Millicom was already a registered company by 1991, they were ready on ground to operate hence they didn't waste time in launching the ETACs mobile standard with the Mobitel brand. Factors such as portability and relative ease of use of the mobile telephones made the mobile service provided by Millicom attractive to urban dwellers. Secondly, most urban dwellers in Accra, Kumasi and some regional capitals were frustrated with their inability to acquire fixed line telephony from the incumbent due to the long bureaucratic process a potential subscriber had to go through to get the fixed-line. For those who had the fixed-line, the quality of service was poor and international tariff was high. Hence Mobitel was on its way to becoming the company with the larger market share. In their first year of operation they had 190000 subscribers (Ghana Web, 2005).

Kludjeson international launched the CellTell brand in 1995. They offered the AMPS mobile Standard. It is not clear why CellTell was not successful in the market, one plausible reason could be that Millicom had a better financial capability than the indigenous Kludjeson International, hence hampering CellTells network expansion.

Millicom's real competitor, Scancom, began operations in 1996. Scancom launched the Spacefon brand delivered via the GSM-900 standard. GSM, a 2G standard was superior to the analogue mobile in the Ghanaian market. The

GSM is digital, which implied that Scancom extend coverage to larger area. They introduced the post-paid subscription (business model competition). The GSM also introduced Ghanaians to personalized functionalities and the call ID functions. Scancom also had superior customer service support to the other mobile networks. Most importantly rather than spend hours talking on mobile phones, you could send an SMS. To bring the service closer to the people, Scancom and Mobitel had mobile kiosks, where one could top-up their units, call friends etc.

By 1997, when the National Communication Authority (NCA), the regulator was established, Scancom had 5000 subscribers roughly a year after they began operations. By that same year, CellTel by 2007 known as Kasapa had 2575 customers. Mobitel still had the lead but it was a big leap for Scancom (NCA, 2013).

The transformation of the mobile telephony market in Ghana unfortunately didn't start as a competitive market. Millicom before the entrance of Kludjeson was a mobile telephony monopoly. As the market evolved, one would say that for a time Scancom had monopoly over the development of GSM in Ghana. Both sides did exploit their monopolies. Scancom did so through the high cost of accessing SIMs. Millicom did so through the cost customers paid to acquire the customer premise equipment (mobile handsets). The cost of Spacefon's Sim card was more than US$80 at the onset. Although Spacefon had these independent kiosk operators in which some would call telecentres, the average Ghanaian could not afford the mobile handset and the SIM pack module. However today, there is a full-fledged competition in the delivery of mobile infrastructure and service of the same or similar standards.

The advent of a competitive market in Ghana, exposed a few problems. The most prominent were interconnectivity problems. These new mobile operators had to interconnect with Ghana Telecoms (incumbent) for international calls and to fixed-line telephone networks. There were complaints from the networks of GT trying to sabotage them. However GT at that time lacked adequate interconnecting infrastructure to accommodate the radical growth of MNO' such as Scancom (Spacefon) (Ajao, 2004). This interconnectivity problems led to higher cost of accessing SIMs as there was more demand than supply. Scancom was able to still attract subscribers, despite this high SIM cost, by introducing the pre-pay business model. There was also innovation in the billing system from the per-minute billing system to the per-second billing system. In this case, although the tariff was still high, the user could control voice call expenditure.

On the technical side, quality of service, although a little better that what the fixed line telephony provided still was poor, especially with SMS. There were latency issues and the problem was due to the fact that the MNO exceeded their subscriber capacity and were expanding to accommodate more.

The pre regulatory telecoms industry suffered from the weak institutional supervision of the industry.

2.3.2 Post - Regulator Mobile Market in Ghana

The regulator, NCA, was established in 1997, three years after the ICT for Accelerated Development Policy was launched. The launch of the Ghana ICT for Accelerated Development Policy (ICT4AD) in 1994 is regarded as the starting point of telecom liberalization in Ghana (Frempong G., 2002). Second bullet in section 1.4 of the policy denotes the intent to establish an open and liberal market economy in Ghana (NCA a, 2003). It was in the National Telecom Policy of 2005, that a clearer picture of this intent to liberalize the telecoms market was made. This policy provided the framework for liberation of mobile telephony along with other telecom network and services. This policy also established a framework for establishing a national regulator and other agencies (GTP, 2013).

Although the mobile market was in its infancy, the national priority of the regulator was the privatization of the incumbent national carrier, Ghana Telecom (GT) and gateway liberalization. The Ghana National Petroleum Corporation (GNPC) led consortium was granted license as the SNO to operate as WESTEL. Both GT and WESTEL were granted mobile licenses in addition to their existing fixed-line license. GT had been partially privatized twice. The first was the management contract and the divestiture of 30% shares to G-Com, a consortium led by Telekom Malaysia. This was a five year contract. The second effort was with Telenor which also ended with little results. The political interferences at the end of each contract led to the constant search for a new investor. Since this book is about mobile development, much discussion will not go into the dynamics of the management contracts as they were geared towards the development of fixed-lines. However, the blessing in disguise from these privatization effort was the transformation of GT into a small conglomerate with different companies. One of these companies which is of interest to us is OneTouch, the mobile telephone arm of the then Ghana Telecoms.

Onetouch, was established in 2001 to deliver GSM services to Ghanaians. It was the same year that Millicom began to develop its GSM network with the brand name Buzz. OneTouch had the advantage of GT's incumbency to provide better QOS for voice calls. OneTouch also had a perfect interconnect with the fixed line telephony system (Ajao, 2004). Based on this advantage they did not compete on the cost of subscriber access as the cost of GT SIM cards were as high as 750 000 cedi (GH cedi 75) an equivalent of approximately US$33 (today's rate). OneTouch had more demand for their services due to QOS but the supply was low due to their limited network capacity. In order to compete, the private network operators had to reduce their cost of SIMs and tariff (voice and data). The limited capacity of OneTouch worked to the advantage of Scancom and Millicom as they expanded their networks. Hence would be subscribers of OneTouch had no alternative than to subscribe to the cheaper SIM and services of their competitors.

Unfortunately WESTEL before it was divested to new owners wasn't successful. The MNOs had no alternative to GT as WESTEL was not functional. WESTEL had its own problems. One of which one would say; the GNPC had nothing to do with commercial telecommunications. That is debatable. However WESTEL was not able to meet the minimum number of fixed-line subscribers stipulated in their license in the first year. Hence they were to pay US$ 25 million penalty to NCA. WESTEL accumulated huge interest as it could not pay the penalty on time. The penalty became a burden to them hence there was no mobile development for WESTEL in the early years.

On the regulatory aspect, the focus of the regulator was on spectrum management and licensing. They also played an arbitration role especially in the early interconnection wars, although not very successful, they tried their best with the resources they had. One could say they exhibited regulatory forbearance and did not provide a tough regulation unless very necessary. In a sense, this is a positive as over-regulation has the potential of stifling competition in a market. NCA also adopted a technology neutral policy to the development of mobile in Ghana. This gave room for internet service providers like Iburst and a few to provide WiMAX services in Ghana to their clients before MNOs could think of 3G.

2.3.3 The Ghanaian Mobile Telephony Market Today

If one would look at it, critically, despite the hiccup start of the mobile market in Ghana the market today has been shaped by more private sector investment, public and private infrastructure development initiatives and competitive business modeling.

- Private sector investment: Unfortunately the amount of investment into the Ghanaian Mobile market was not available at press time. However the market attracted more investments which came via corporate take overs and divestitures. Scancom took the lead with takeover by Investcom (owners of Areeba brand). Subsequent majority of Investcom was acquired by the South African Mobile Telephone Network (MTN). Today its brand is the MTN brand. Millicom had to up its stakes in the investment and expansion of its networks with its new brand Tigo. WESTEL was bought by ZAIN and now owned by Bharti Airtel with the brand Airtel. Kludjeson went through a divestiture where Hutchson bought 80% stakes and later Expresso Telecoms, owned by Sudatel bought a 100% stake of the company, hence its new brand name Expresso. Vodafone now holds 70% stake of GT conglomerate and there is a new mobile network from Nigeria, GLO.
- Public and Private Infrastructure development: There has been expansion of the backhaul fibre-optic network infrastructure of these companies. Most of the infrastructure are the MNOs initiative. The government of Ghana has also invested money in this regard as well. The competition in infrastructure development has been enabled by fear and not as a result of deliberate need to improve infrastructure. The fear of interconnecting with a rival MNO led to fibre-optic backhaul development. The fear of being kicked out of the market by the new entrant with 3G technology also led to the development of 3G in the country. This happened in the case of ZAIN and GLO's entry into the market.
- Competitive business modelling: A little has been mentioned about this so far on the pricing mechanism. However other initiatives such as, Game shows, music shows, sponsorship of events, giving out free SIM cards and many other incentives have been used to attract more customers to their networks.

Today a brief snapshot of the market reveals the following. By September 2013 the market share of each MNO was as follows: Scancom (MTN) 46.33%, Vodafone 21.13%, Millicom (Tigo) 13.89%, Airtel 12.21%, GLO 5.87% and Expresso recorded 0.56% of the Market share.

2.3.4 Conclusion

In the Ghanaian case it is clear that liberalization enabled competition not only in the delivery of services but also in the development of infrastructure and competition of business models. Although the institutions that governed the industry were weak initially, one would say that gradually they are coming up. However, the regulatory forbearance practiced by the NCA in a way has not stifled the market. As mentioned earlier, this act of regulatory forbearance has not come without consequence as Expresso seems not to be protected. Expresso has lost out, at least for now, of the competition battle in the mobile market. However it is important to note that Expresso is doing well as an Internet Service provider. Ghana Telecoms, one could say was lucky to be salvaged by Vodafone, thereby expanding its capacity and putting it to the part of competitiveness.

The sad story though is the lack of indigenous participation in this market and the inability to actually measure in real terms the number of mobile phones per house hold. Although the penetration rate is above 100%, many villages in Ghana have no access to mobile telephony and some innovative regulation or market intervention is needed to push the development of the mobile to the interior.

It is worth mentioning that although liberation played a major role in the development of the mobile telephony market, other factors such regulation of right of way and has enabled the modest development of competitive fibre optics backhaul in Ghana. Public investment (e.g. Voltacom fibre optics) served as a market incentive to extending mobile coverage to most parts of Ghana. The liberalization of international bandwidth as well played a great role towards tariff reduction. The price for an E1 connection due to international bandwidth liberalization reduced from US $ 4500 to US $1200. This one would say has also made an impact in the pricing of telecom services in Ghana.

Case 2: Nigeria

Nigeria is the most populous country in Africa. Nigeria was also one of the first countries to pass the market reform law in Africa, however the implementation of this law did not really materialize until 2001 in the start of this century. In the case of Ghana, the advantage they had was the prolonged rule of Jerry John Rawlings and his PNDC and later NDC party. In the case of Nigeria, although the Law was passed by the Ibrahim Babangida military junta, it was not until the restoration of democracy in 1999 that something had to be done about implementing the reform. The only aspect of the reform that occurred earlier was the commercialization of MTEL (the mobile arm of NITEL, the national monopoly). MTEL was established in 1996 by the Federal Government of Nigeria (Motorola). MTEL inherited the analogue mobile systems of NITEL and begun operations as a separate government owned entity.

In 2001, MTEL was merged once again with NITEL and acquired a GSM License. Other winners of the license auction were Econet Nigeria ltd, MTN Nigeria and CIL. Unfortunately CIL lost its license.

The development of the mobile market in Nigeria was rapid due to the fact that market was huge and operators wanted to capture as much of the market as possible. The MNO's at the initial stage, competed in the development of GSM infrastructure to Nigerian cities. There were a lot of branding effort aimed at promoting the various mobile packages. MTN led the pack. The advantage they had was that their closest rival Econet were entangled with shareholder and management problems. MTEL wasn't in the best shape to totally compete as it was part of the initial failed efforts in the privatization of NITEL.

At the inception of the liberalized mobile market in Nigeria, demand for mobile telephony was not very high due to the high cost of SIM cards and high tariff (voice and data). Subscribers were billed per-minute and it wasn't very economical to have a proper conversation on mobile telephones. MTN saw no need to change from the per-minute billing to the per-second billing as they has a greater infrastructure penetration than the rest. The MNOs capitalized on NITEL's poor fixed line services to deliver their services. It was in 2003, a year after that the National Communications Commission (NCC), the Nigerian telecoms regulator liberalized the Gateway. Global communications (GLO) an indigenous Nigerian was granted the SNO license as well as a GSM license. GlO reduced interconnection fees for international calls and also introduced the per-second billing system. GlO also gave out their SIMs for 1 naira compared to more than 200 naira charged by MTN and the rest. This act forced MTN and Vmobile (former Econet and now Airtel) to reduce their voice and data tariff. MTN was also forced to adopt the per-second billing which they initially said was impossible and Vmobile was forced to look for an investor (ZAIN, nor Airtel) to help them compete.

The liberalization of mobile telephony in Nigeria has also resulted in huge financial investment in the sector. By the end of 2002 before GLO commenced operations. The Nigerian telecoms market was worth US $ 1.1 Billion with an annual growth of 37% (Wills & Daniels, 2003). MTN was (still is) the highest investor in infrastructure in 2002 with US $ 1.4 Billion and the industry Average Revenue per User (APRU) was US $ 54 (Wills & Daniels, 2003). Today, the total investment in the telecoms sector in Nigeria from the inception of liberalization till June 2013 was estimated at US $ 32 Billion (Akwaja, 2013). Between 2002 and 2008, Nigeria reaped US $ 2.2 Billion from Spectrum Licensing (Mobile News, 2008). Between 2011 and

2013 US $ 7 Billion, was spent in infrastructure development by the leading 4 telecom Giants MTN, Airtel (Former Vmobile), Etisalat, and GLO. Due to the ensuing completion in the market the ARPU in 2012 was 949 Naira (US $ 5.9) (Okwuke, 2012). The low ARPU is attributed to usage of value added services by Nigerians.

Telecom market liberalization has led to an increase to Nigeria's Teledensity. Thanks to mobile telephony, the teledensity increased from 0.73 in 2001 with 266, 461 GSM mobile lines to 80.85 with 135 Million GSM lines and 14 Million CDMA lines in 2012 aside fixed wired and fixed wireless line services (NCC, 2013). By the end of 2013, The GSM operators MTN had more than 47 Million subscribers, GLO had more than 24 million subscribers, Airtel had more than 23 Million subscribers, Etisalat recorded more than 14 million subscribers, MTEL recorded a little above 250,000 subscribers. Among the CDMA operators, Visafone recorded a little above 2 million subscribers, Starcomms, Multilinks and Reliance Telecom recorded between about 100, 000 subscribers to 300,000 subscribers. There are other companies in Nigeria providing wireless services with IP Telephony, WiMAX and other Broadband Wireless Access (BWA) Services.

On the aspect of regulation, MNOs are obligated to provide QOS else they are fined. The NCC also granted unified licenses to small telecom companies with Universal Service Obligations. The obligation of the unified licensing was the extension of the service of the operators to the stated of the federation. Companies like Multi-Link and Starcomms who were fixed line telephony Operators and Internet Service providers in Lagos and Port Harcourt had to extend mobile telephone services (CDMA) to many parts of the country. Although the unified licensing enabled the small MNOs to extend their Services from Lagos, Port Harcourt and Abuja, the spread of mobile telephony was also characterized by the growth of the economy of various Nigerian States. This led MNOs to establish their services in state capitals and later in local government areas.

The National Communications Commission (NCC) was poised to establish its foot hold as a firm referee. They had a hands on approach to monitoring the market and the services delivered. One way of monitoring quality of service and ensuring that customer complaints were met was the establishment of the customer parliament. This parliament was chaired by the head of the regulatory agency and representatives of the mobile telecom companies were present. This was aired on national television and people brought various complaints and the regulator made sure the problems were solved or the company was sanctioned.

The growth of the Nigerian Mobile market has not been without challenges. These include regulatory challenges which have led the regulator to develop regulations on interconnection, frequency pricing, Quality of Service, Enforcement procedures, Universal Access and Service, Code of practice, competition practice and numbering (NCC). One would not say that these regulations are often enforced as there is a balance between over-regulation and under-regulation of the industry in most cases through penalties and levies.

The Nigerian mobile market is one of the fastest growing markets with a huge population of potential subscribers to reach out to. Infrastructure development has been the prerogative of the private sector.

Nigeria and Ghana have the similar problems of individuals owning more than one mobile telephones. Hence it is difficult to access the actual penetration of mobile telephony in Nigeria. The Nigerian business environment is tough and very unpredictable. This in itself is a challenge, however, despite this challenge, the liberalization of the market has led to affordable mobile services, innovation in service delivery, there is competition in the delivery of 3G and LTE (provided by GLO) and quite unlike many other African countries, there are a lot of local mobile companies operating in Nigeria providing different brand of mobile services.

Case 3: Kenya

The Kenyan mobile telephony market may not have attained the level of penetration experienced in Ghana, neither would one say that it is as huge like the Nigerian market. However, the Kenyan mobile market is dynamic for one clear reason. The privatization of Kenya Telkom gave birth to Safaricom and Orange (both mobile service providers) which has accounted for more than half the mobile subscribers in Kenya.

The evolution of Safaricom began with Kenyan Post and Telecommunications Corporation (KPTC) who provided fixed-line telephony before the telecom market liberalization. KPTC created Safaricom as the mobile arm of the P&T in 1993. They provided the ETAC mobile Standard to their customers. Safaricom upgraded their network GSM in 1996 but they got the GSM license in 1999. By 1997, Safaricom was incorporated as a private Limited Liability company in 1997. In 2002 as a way of injecting more management expertise and financial muscle to the company, the shares of the company was offered to the general public. The Government of Kenya held 60% shares through Telkom Kenya. Between 1997 and 2000, Safaricom's subscriber base grew from 3000 subscribers to 54000 subscribers (Kane, 2002). Today, Safaricom has a subscriber base of more than 17 million customers. Safaricom's growth was as a result of Kenyans being frustrated with the poor fixed-line services provided by KPTC.

In many African countries, once the mobile arm of the Incumbent monopoly decoupled from the parent network operator, they provided only fixed-line services. In Kenya the case was different. The sector deregulation in 1999 led to the decoupling of the telecoms arm of KPTC from the Postal arm and the regulator was created (Kane, 2002). The Communications Commission of Kenya (CCK) was formed by the Kenya Communications Act No.2 of 1998. Under this act the CCK regulated both the telecom and postal service (CCK, 2013). The duties of the regulator changed to regulating only telecommunications by the amendment of the Act by 2009. The new Incumbent monopoly was called Telkom Kenya under the company's act 1999 (Telkom Kenya, 2013).

The takeoff of Telkom Kenya as a single company was not smooth as it was riddled with debt and inefficiency. However the International Finance Corporation (IFC) helped secure US $ 81million to enable the downsizing and employee transition of Telkom Kenya (PIDG, 2013). Telkom Kenya pledged part of its 60% stake in Safaricom, as collateral for the capital raised by IFC. The loan was to be repaid from the proceeds of the privatization of Telkom Kenya. The essence of the downsizing was to enable Telkom Kenya attract investors. This process led to the later inevitable unbundling of Safaricom from Telkom Kenya where 25% of the Telkom Kenya share in Safaricom was issued to the public (IFC, 2013). Hence Telkom Kenya had no controlling stakes in Safaricom any longer (Safaricom, 2013). 51% shares of Telkom Kenya was won by France Telekom at an auction facilitated by IFC in 2007. French Telecom upgraded their stakes to 70%. Today Telkom Kenya through the Orange brand is competing as a different company with Safaricom.

Aside the result of privatization, there are two private players in the mobile market, Airtel and Yu Mobile. Kencell, now Airtel Kenya was the second company after Safaricom to be awarded 2G license in Kenya (Sameer Group, 2013). 60% of Kencell was owned by an indigenous company Sameer group (Kane, 2002). The remaining 40% shares were owned by French partners Vivendi. Zain acquired the stakes of vivendi and eventually, as Airtel acquired, the African arm of ZAIN, The Company was rebranded Airtel. In their first year of operation providing GSM, they recorded 60000 subscribers (Kane, 2002). The entrance of Kencell to the market led Safaricom to reduce the activation fee for subscribers from Kshs 10000 to Kshs 2000 in 2002. The tariff per minute of Kshs 28 before 1999 was reduced to 10Kshs in 2002 (Kane, 2002). In 2008, Yu Mobile owned by Essar became the 4th GSM operator in Kenya. This led to competition, private capital injection and Government interventions (via the building of International bandwidth capacity) which has resulted in reduced tariff charges. In 2008 Yu Mobile charged Kshs 7.5 per minute to other networks and customers were given the incentive of Kshs 0.75 for receiving calls from other networks.

Despite the relative growth in the mobile telephony market, the penetration of mobile telephony in rural Kenya is low. To ensure rural telecom infrastructure development, the 4G LTE license will be granted to the four incumbent operators and other new infrastructure providers and telecom operators as a group. The nine Member consortium involving both incumbent and new operators include, Kenyan Government treasury, Safaricom, MTN, Airtel, Essar, Telkom Kenya (Orange), KDN, Alcatel- Lucent and Epesi formed a consortium. The essence of the consortium is to enhance infrastructure sharing, thereby lowering the cost of the telecom infrastructure. By this principle, the cost of investing in rural areas will be cheaper. The Public Private Partnership approach to LTE development is laudable but there seem to be some reluctance in the part of some telecom companies who would want to deploy their network by themselves.

2.4 Conclusion

The results of liberalization have been positive and challenging as well. The positives as seen in the cases include: The lowering of market entry requirements and in some cases tax incentives have led to the influx of both international and regional operators into the market. Reduced tariffs has led to affordable access of mobile telephone starter packs. Gateway liberalization, public investment in international and national backhaul infrastructure and national backhaul liberalization has played a supporting role to mobile development. Innovative business models and the provision of innovative value added services has been enabled by the liberalized market. The liberalized market has been the catalyst to the development of mobile broadband standards.

The challenges include regulatory challenges fostered by technology convergence. There is also a challenge of political interventions, especially in former public owned companies. The biggest problem the regulators in Africa face is drawing a balance between the provision of mobile as a mixed good (Part public/part common good), and the provision of mobile in a way the operator will make reasonable profits. In some African countries, the mobile network operators are seen as cash cows. Hence this becomes a challenge as the MNO tries to compete and also make a decent profit.

Although the mobile telephone market in Africa is growing, Mobile broadband isn't growing that fast. Aside from that, most of rural Africa has little or no telecom infrastructure. These areas are commercially unviable, hence there is need for some interventions to help push the mobile network frontier forward. Having accessed the impact of liberalization on the market, the question now remains: is there a need for another market reform mechanism to help push mobile into rural Africa.

3

Telecommunications in Africa – Regulation, Technologies, and Markets

Godfred Frempong
Council for Scientific and Industrial Research
Science and Technology Policy Research Institute
(CSIR-STEPRI) Ghana
gkfrempong@csir-stepri.org

Anders Henten
CMI, Department of Electronic System
Aalborg University, Copenhagen, Denmark
henten@cmi.aau.dk

3.1 Introduction

The aim of this chapter is to provide a compressed overview of the developments and status of telecommunications in Africa and a selection of African countries. The purpose is not to provide an in-depth analysis of these developments but to present the most salient overall facts. Emphasis is on mobile, as this is the technology, which has contributed most effectively to the spread of telecommunications in African countries. Prior to the development and deployment of mobile technologies, telecommunications were in an under developed state in most African countries. The costs of deploying fixed-line telephony were simply too high compared to the general income levels, and the business models and regulatory frameworks were not conducive for any wide-spread developments of telecommunications.

First, there is a brief overview of salient regulatory trends in Africa: How, when, and to what extent have African countries implemented the general international liberalization of telecommunications? This is followed by an equally brief overview of technology trends in Africa including fixed telephony, mobile and wireless, Internet, and broadband.Thereafter, the current status of the main access technologies is presented based on figures provided by the ITU, followed finally by a short summary.

3.2 Telecommunications Regulatory Trends

In telecommunications, regulation has become a core issue to ensure an effective development of the industry globally and within national economies. Regulation means not only the provision of rules and regulations and exacting obedience from the players of the industry. It has also become a necessary mechanism to attract investments into the industry. Consequently, regulation plays a dual role of ensuring sanity in the markets as well as serving as catalyst to attract investments.

A significant development in the regulation of telecommunications was the conclusion of the agreement on basic telecommunications services under the World Trade Organization (WTO) in 1997. This is an international agreement on the liberalization of basic telecommunications and it marked an important milestone in the telecommunications industry's shift towards global competition and open markets. African countries which have joined the WTO, have had to sign and adhere to the tenets of the Basic Telecoms Agreement. As at June 2013, 42 African countries were members of the WTO.

African countries have generally liberalized their telecommunications market. The liberalization has taken place not only in voice telephony but also in data and multimedia services. Most African countries have more than two mobile companies. For example, Kenya has four mobile network operators; Ghana has six, Senegal three, Nigeria nine, and Egypt three. This has contributed to the rapid growth of the mobile markets in Africa and to creating dynamic markets with competitive pricing.

In the fixed telephone markets, partial liberalization in terms of duopoly has been introduced to bolster competition. However, the explosion seen in mobile technology cannot be witnessed in the fixed area. Fixed telecommunications is only developing slowly in Africa.

Most African governments have privatized the state-owned operators by selling part of the shares to strategic investors which mostly were foreign-based telecommunications operators. For example, France Telecom has bought shares in Sonatel in Senegal and in Telkom Kenya. Ghana initially sold 70% of the then Ghana Telecom to Telkom Malaysia in the 1990s before it was resold to Vodafone of United Kingdom in 2008. The rationale behind the privatization has been to attract foreign capital and expertise to invigorate the underperforming public network operators.

In addition to the liberalization of the telecommunications industry, many African countries have introduced innovative licensing regimes with the aim of fostering development in the markets. There is a trend of introducing universal/unified licensing regimes in the African markets. Unified license regimes have been introduced in Nigeria, Kenya, Tanzania and Uganda among others. The implication is that an operator with a unified license can operate in different market segments provided it has capital and the technical capability to do so (Frempong, 2011).

Thus, regulatory progress has been made. However, there are still many issues to be resolved. Figure 3.1 provides information of the findings of an assessment of the telecommunications regulatory environment (TRE) in selected African countries conducted by Research ICT Africa in 2011and 2012. The assessment is based on opinions of key stakeholders in the telecommunications industry and covers fixed-line telephony, mobile telephony and broadband. The assessment is based on seven indicators.

The figure shows that all selected countries have had a negative score with respect to the regulatory environments. However, countries with the worst scores were Ethiopia, South Africa, and Mozambique. The implication of the assessment is that African regulatory institutions need to work extra hard to develop conducive environments to support telecommunications developments

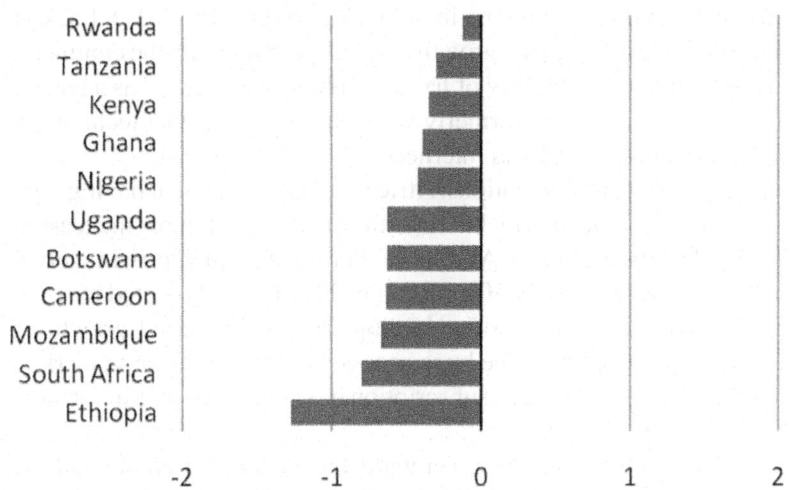

Figure 3.1 Overall TRE scores in 11 African countries

Source: Frempong, 2012

From this brief overview, it is clear that governments in Africa have responded to the regulatory challenges in the global telecommunications industry. Markets have been liberalized and initiatives have been implemented in order to introduce dynamism into the markets. However, the assessment of the telecommunications regulatory environments indicates that much still remains to be done to ensure regulatory proficiency in Africa.

3.3 Telecommunications Technology Deployment

Before the 1990s, the telecommunications services available in Africa were basic telephony, telegram, and telex. These services were provided largely by state-owned telephone operators. The deployment of these services was limited to commercial and urban towns. Technological developments in microelectronics, however, resulted in the development and deployment of many telecommunications technologies, which contributed to changing the landscape of the global industry including those in Africa.

Fixed telephony has a long history in Africa. It dates back to the colonial era, where the services (voice telephony and telegraph) were introduced to link commercial centers to facilitate trade and governance (Frempong, 2007). In the post-independence era, fixed telephony continued to dominate the African communication sector until the 2000 decade, where its dominance was reduced drastically. In 2010, the penetration rate of fixed telephones in Cote d'Ivoire was 1%, Benin 2% and less than 10% in South Africa (GSMA, 2011). Lack of investment negatively affected the growth of fixed telephones on the continent. The poor growth and unavailability of fixed telephones has served as a barrier to the uptake of services which formerly were only based on that technology (for example, fixed broadband and internet).

In terms of mobile communications, Africa has been taunted of having high growth in penetration in the world, but in terms of absolute figures the least in the world. Mobile subscription in Africa has increased from approximately 5 million in 2000 to approximately 500 million in 2010 and is expected to reach approximately 800 million by 2015 (Harding, 2011). The rapid growth of mobile telephony has addressed the huge access gap in the telephone market. The novelty, flexibility, and ease of subscription have provided a fillip for such growth.

In terms of technology, analogue networks have been phased out and the networks are now dominated by GSM and CDMA. The deployment of 3G is evolving in Africa. By mid-2011 only about 15 countries in Africa had not deployed 3G infrastructure (ITU, 2011). This shows that African mobile operators are adopting modern technologies to provide services. However, coverage is still lacking in large areas. For example, 40% of the MTN network in Ghana consists of 3G technology but is concentrated in the main cities of the country. Large portions of the country remain uncovered by 3G.

The growth and penetration of mobile telephony in Africa has led to the development of innovative applications such as mobile money transfer, infrastructure to access important information on market prices, weather, and authenticity of drugs among others. Mobile money services have gained much credence in countries such as Kenya, Tanzania and Uganda. It has become a big business in these countries. In 2010, for example, 14 million users of M-Pesa in Kenya transferred an estimated USD7 billion and this was equivalent to 20% of the country's gross domestic product (Anchan, 2012).

Internet services became commercially available in Africa in the late 1990s. Initially, the service was a basic dial-up system with a speed of 256kbps. However, the development of broadband internet has replaced the dial-up system. But internet service providers, at first, relied on the network of the fixed telephone operators, and the poor penetration and quality of the fixed telephone networks are among the factors that have affected the penetration of internet and broadband services in Africa. For example, 167 million people out of the total population of one billion in Africa have access to the internet and broadband services (Internet Society,n.d.). However, connectivity according to the Internet Society is growing in Africa due to the increasing deployment of mobile broadband services.

The technological development in the telecommunications industry and the poor performance of the fixed telephone networks have provided an increasing demand for wireless technology, either as the last mile or as the main infrastructure for internet and broadband services. Consequently, wireless infrastructures such as 3G, Worldwide Interoperability for Microwave Access (WiMAX) and Long Term Evolution (LTEs) are complementing and, in most cases, substituting the development of cable and DSL. A number of countries have adopted WiMAX for data services and LTE services are developing in Africa. The bandwidth capacity of LTE may have an advantage in its deployment. LTE technology in the current version can achieve download speeds of about 100Mbps, while that of WiMAX is around 40 Mbps.

The development and deployment of LTE as the next generation mobile broadband and internet infrastructure is evolving. African countries that have commercially deployed LTE services include Angola, South Africa, Namibia, Mauritius, and Uganda (Global mobile Suppliers Association, 2013). The operators that have launched this service are Movicel in Angola, Orange in

Mauritius, Namibia's MTC, Smile in Uganda and MTN and Vodacom in South Africa.

One problem which may hinder the rapid deployment of LTE network in Africa is investment. Many mobile operating companies have invested heavily in deploying 3G technology networks, and may, therefore, be hesitant to add investments in LTE networks. This is a critical issue since many mobile operators have not completed upgrading their network from 2G to 3G.

On the continental scale, there has been a proliferation of submarine fiber cables in Africa. Currently there are about 10 submarine cables connecting different parts of Africa to the rest of the world. Mention can be made of LION 1 &2, ACE, SEAS, SAT3, WACS, SEACOM, and SEA-ME-WE 3&4. This is in contrast to the situation in 2000/2001 when SEA-ME-WE3 and SAT3 were the only cables available in Africa (Young, n.d.). The existence of these submarine cables provides a foundation for fast internet and broadband connections between Africa and the rest of the world.

To provide communications, most African countries have constructed terrestrial backbone fiber systems. Mention can be made of Ghana, Nigeria, South Africa, Namibia, Botswana, and Zambia as some of the countries with active national fiber backbone systems. The total national fiber backbones in Africa were recently at 615,000 km (Internet Society, n.d.) and are continuously growing in order to provide good national infrastructure for fast internet connectivity.

Africa has thus kept pace with the developments in telecommunications technologies. This applies mainly to the mobile field, where the newest technologies are relatively quickly being implemented. In the fixed area with technology solutions building on existing technologies, for instance DSL on the top of PSTN, there is little development, while fiber technology is used in the backbone and only to a very small extent in the access paths.

3.4 Market Trends

International Telecommunication Union (ITU) reports on the developments regarding access technologies, internet use, and broadband penetration in the publication 'Measuring the Information Society 2013' (ITU, 2013). In this section, figures for the present status (2012) are presented for a selected 15 African countries representing those that are the most developed in terms of telecommunications and those that are the least developed.

The overall status (2012) shows that[1]:

- An estimated 16.3% of individuals use internet in Africa
- An estimated 10.9% of Africans have an active mobile broadband subscription
- An estimated 0.3% of Africans have a fixed broadband subscription
- An estimated 6.7% of African households have internet access

These figures are averages for all African countries. As we shall see in the more detailed statistics, there are considerable uncertainties with respect to the figures for a large number of African countries, which obviously affects the average figures. They, nevertheless, represent a best estimate[2].

In the remainder of the section, figures for a selection of Sub-Saharan countries will be presented concerning the following categories:

- Fixedtelephone subscriptions per 100 inhabitants
- Mobilecellular subscriptions per 100 inhabitants
- Percentageof households with internet access
- Percentage of individuals using the internet
- Fixed broadband subscriptions per 100 inhabitants
- Activemobile broadband subscriptions per 100 inhabitants

The sequence of countries in the tables is determined by their degree of performance with respect to the ITU ICT Development Index (IDI).

The figures in Table 3.1 document a very low penetration of fixed telephone subscriptions in general. Even the countries with the highest take-up rate, South Africa and Botswana, have a relatively low penetration of less than 10%. And, in most other countries, the penetration rate is around 1% - which reflects the underdeveloped state of fixed-line communications from the colonial past and the post-colonial era. After the liberalization starting in the 1990s, however, mobile has developed faster. In an increasing number of African countries, the penetration rate is already above 100 percent reflecting a widespread take-up of mobile communications. It also reflects the fact that many subscribers have more than one subscription. There are still considerable areas and population segments, which are not covered or cannot afford mobile communications.

Table 3.2 illustrates the development of internet use, which obviously depends on the access paths reported on in Table 3.1.

---

[1]Summary by Africa – Tracking Internet Progress, www.oafrica.com, retrieved 02-01-2014.

[2]The ITU figures, however, differ at important instances from figures published by Research ICT Africa for 11 African countries for 2011/12.

Table 3.1 Fixed telephone subscriptions and mobile cellular subscriptions per 100 inhabitants, 2012

	Fixed Telephone Subscriptions	Mobile Cellular Subscriptions
South Africa	7.9	134.8
Botswana	7.8	150.1
Ghana	1.1	100.3
Kenya	0.6	71.9
Nigeria	0.3	67.7
Senegal	2.6	87.5
Uganda	0.9	45.9
Cameroun	3.6	64.0
Angola	1.5	48.6
Tanzania	0.4	57.1
Malawi	1.4	27.8
Mozambique	0.4	33.1
Ethiopia	0.9	23.7
Burkina Faso	0.8	57.1
Niger	0.6	32.4

Source: ITU: Measuring the Information Society 2013, pp. 228–229

Table 3.2 Percentage of households with internet access and percentage of individuals using the internet, 2012

	Percentage of Households With Internet Access	Percentage of Individuals Using The Internet
South Africa	25.5	41.0
Botswana	9.1	11.5
Ghana	11.0	17.1
Kenya	11.5	32.1
Nigeria	9.1	32.9
Senegal	5.8	19.2
Uganda	4.2	14.7
Cameroun	3.5	5.7
Angola	7.2	16.9
Tanzania	5.1	13.1
Malawi	5.5	4.4
Mozambique	4.7	4.8
Ethiopia	1.9	1.5
Burkina Faso	2.8	3.7
Niger	1.4	1.4

Source: ITU: Measuring the Information Society 2013, pp. 228–231

The figures in Table 3.2 are subject to a greater statistical uncertainty than the figures in Table 3.1. However, the figures illustrate that a relatively large percentage of the populations in African countries uses internet. In South

Table 3.3 Fixed broadband subscriptions and active mobile broadband subscriptions per 100 inhabitants, 2012

	Fixed Broadband Subscriptions	Active Mobile Broadband Subscriptions
South Africa	2.2	26.0
Botswana	0.8	17.4
Ghana	0.3	33.7
Kenya	0.1	2.2
Nigeria	0.0	18.6
Senegal	0.7	3.8
Uganda	0.1	7.6
Cameroun	0.0	0.0
Angola	0.2	1.5
Tanzania	0.0	1.5
Malawi	0.0	3.5
Mozambique	0.1	1.8
Ethiopia	0.0	0.4
Burkina Faso	0.1	0.0
Niger	0.0	0.6

Source: ITU: Measuring the Information Society 2013, pp. 230–231

Africa, more than 40 percent of the population used internet in 2012[3]. In other countries, which are relatively advanced in terms of telecommunications, for instance Kenya, a third of the population uses internet. And, these figures are growing quickly. Household access is primarily via mobile connections, and the figures for individual use are generally higher than household access, as there are also possibilities, for instance, for internet access at internet cafes or using other people's access lines. However, the figures must beconsidered as relatively uncertain, for example in the case of Malawi, where household access is higher than individual use.

In Table 3.3, broadband subscription is depicted. As can be seen from the figures in the table, these are rather uncertain in several cases.

In spite of the uncertainty, for instance that the percentage of active mobile broadband subscriptions is higher than the percentage of individuals using internet in Botswana as well as Ghana, the figures illustrate that broadband access is almost entirely taking place by means of mobile broadband. Mobile

[3]The data contradicts the findings of a survey carried out by Research ICT Africa in 11 African countries in 2011/12. From the survey, household internet was found to be only 0.8 percent in Tanzania, 1.3 percent in Cameroon, 3% in Ghana and nearly 20 percent in South Africa.

technology is not only a means for voice telephony and SMS but also for broadband access. Fixed broadband access, on the other side, is at a very low level and is not likely to develop much in the coming years. Investment costs in mobile technologies are far lower than in fixed technologies and the costs for fixed broadband users would, consequently, be at an unaffordable level for most users.

3.5 Summary

In the chapter it is shown that African countries in general have followed the overall trends of liberalization, privatization, and regulation seen all around the world. It also shows that African countries are able to keep pace with respect to new telecommunications technologies in the mobile field. The penetration rates, for most African countries, may not be at the same level as in, for instance, European countries. However, mobile communications is developing at a speed that was not envisioned 20 years ago - though there are still considerable parts of the populations, which have no coverage or cannot afford traditional mobile communications.

It is clearly illustrated that access to traditional telecommunications services, such as voice telephony and SMS, as well as internet access takes place via mobile platforms. The underdeveloped state of telecommunications from the colonial era as well as the post-colonial years has meant that the fixed access infrastructure is very scanty, and as mobile infrastructures are less costly to deploy, the present and future of telecommunications in African countries will generally be based on mobile platforms.

4

The Prepaid Mobile Market in Africa

Roslyn Layton
Strand Consult &
CMI, Department of Electronic System
Aalborg University, Copenhagen
Denmark
rl@cmi.aau.dk

4.1 Introduction

Africa is the world's final frontier for telecommunications. While mobile phones and subscriptions have been extant for some time, connecting the continent with broadband represents a key commercial, governmental and social goal. Overall Africa's rate of internet penetration is about half of the world average. To date, of the few broadband connections in Africa, the vast majority are provided by mobile networks. This is likely to be the case going forward given mobile's relative low cost to other networks, the geographic features of the continent, and users' comfort with the technology. Informa, a leading telecom research firm estimates mobile broadband connections to grow to 250 million by the end of 2015, delivering over 20% of its total internet traffic. These projections also include a doubling of smartphone ownership. (Informa Telecoms and Media, 2012) These data are predicated on the assumption that the mobile industry will be able to deliver the necessary infrastructure and related goods and services. This also requires a shift from the prevailing voice and SMS packages offered today to ones that focus on data. Therefore the economics of mobile subscriptions are a valuable area of investigation.

Africa is a continent with over a billion people in 54 countries each with its own history and situation. The data available tends to focus on a few of the larger countries or the region as a whole. This can obscure important differences between countries. However some generalizations can be made.

More than half of the world's mobile services are offered on a prepaid basis. This is a fact that rarely crosses the mind of people in developed Western countries where post-paid mobile contracts dominate. People in the West may use prepaid when on vacation or purchasing a mobile plan for a friend or relative. However in most African countries, prepaid service makes up more than 90% of the market.

When people discuss the mobile miracle in Africa, they tend to focus on mobile phones. Aker and Mbiti have noted the impact to agricultural and labor market efficiency and welfare (2010), and others have touted the benefits for health services delivery, education, microfinance and so on. But the underlying business model—prepaid—is rarely if ever discussed. While phones are important, it may also be that Africans having access to inexpensive mobile subscriptions without having to present financial or identification credentials is part of the explanation for mobile's impact in Africa.

Prepaid offers easy and convenient access to modern communication for all type of customers. It has a number of advantages, but there some areas of the prepaid market in Africa that can be optimized and improved. Mobile operators in Africa are faced with the challenge of evolving their business model from offering voice/SMS packages to data packages. Revenue and ARPU from voice are higher and more profitable than the selling of data. This is driven by a number of factors including technology, consumer demand, and increasing penetration of smartphones. Operators compete not just on price, but also on technology. As such, any improvement that a mobile operator can make to its business practices, can improve the efficacy of achieving the big picture broadband goals.

Consumers eager for internet services will start to buy data packages, and they substitute voice and SMS for over the top (OTT) services. Smartphones are increasingly inexpensive and available. Even feature phones have some capabilities associated with smartphones. Nokia makes a line of phones that has a range of apps and email functionality. In any case, as Africa develops, new web-based services will be available, and the de facto way to access them will be the mobile phone, not the personal computer or laptop. In OECD countries the web has evolved from pc to laptop to mobile device, but in Africa and other developing regions, the web experience is largely mobile first and only.

The goal of this chapter is to review the prepaid market in Africa and its challenges and to suggest ways in which the relevant actors can improve the

Table 4.1 Brief overview of broadband penetration in Africa

The International Telecommunications Union (ITU) compiles data about telecommunications in every country in the world. It prepares a number of reports with different indicators. A brief overview shows that fixed broadband subscriptions have not grown very much, but mobile broadband subscriptions have increased more than 50 fold in just 6 years. While Africa overall has low average broadband penetration compared to other regions, looking at broadband data in each country, shows even further gaps. Just 8 African nations have broadband penetration above 40%: Algeria, Angola, Benin, Botswana, Burkina Faso and Burundi. The average rate of individuals with broadband connections is just 14%. Meanwhile with the exception of Botswana, none of the countries with high GDP per capita in Africa (Equatorial Guinea, Seychelles, Gabon, Mauritius, Libya, South Africa, Botswana, and Tunisia) score in the top 8 for broadband penetration. A chart illustrating the results follows is seen in the appendix below.

Broadband subscriptions by region

Fixed subscriptions:	2007[a]	2010[a]	2013[a,b]
Africa	0.1%	0.2%	0.3%
Americas	10.9%	14.1%	17.1%
Arab States	0.9%	1.9%	3.3%
Asia and Pacific	3.2%	5.5%	7.6%
Commonwealth of Independent States	2.3%	8.2%	13.5%
Europe	18.4%	23.6%	27.0%
Mobile subscriptions:	**2007[a]**	**2010[a]**	**2013[a,b]**
Africa	0.2%	1.8%	10.9%
Americas	6.4%	22.9%	48.0%
Arab States	0.8%	5.1%	18.9%
Asia and Pacific	3.1%	7.4%	22.4%
Commonwealth of Independent States	0.2%	22.3%	46.0%
Europe	14.7%	28.7%	67.5%

[a] Per 100 inhabitants. [b] Estimate.

Source: International Telecommunications Union.[2][1]

Source: ITU

market for the benefits of African consumers. For many operators it may be a matter of reducing costs with a smartphone deployment of prepaid services. Cost reduction, particularly in the area of distribution and top up of prepaid balance can help compensate for the negative impacts of high churn and price decline. A shift from scratch cards to electronic top can represent a 40% cost reduction. Similarly operators in the prepaid market face high churn rates, 3% or more. Addressing these two issues can translate to an EBITDA improvement of 2–3% within 3–6 months. With better margins and more revenue, operators can focus on other activities such as expanding service, upgrading infrastructure and so on.

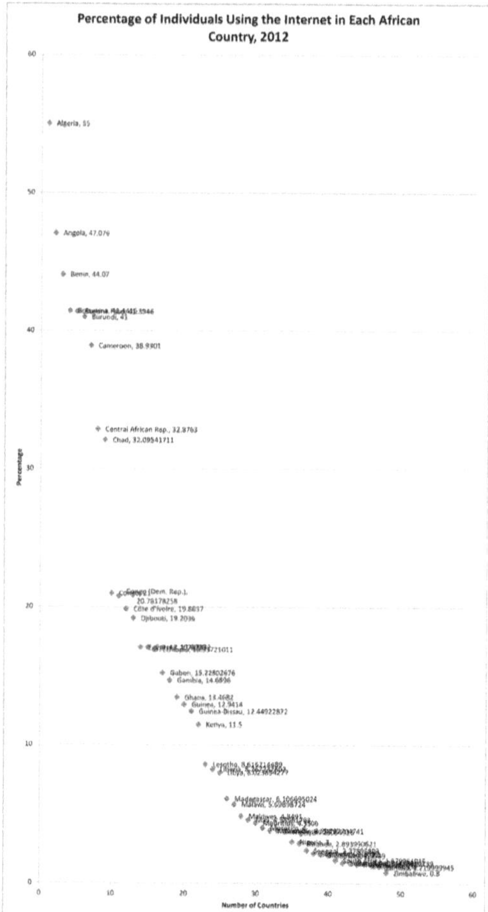

Figure 4.1 Broadband penetration by African country, 2012

Source: ITU

4.2 Advantage and Disadvantages of Prepaid

The first prepaid phone cards were offered in Italy in 1976 as a solution to the shortage of coins in the country and vandalism of phone booths Over the coming years prepaid concepts made inroads to traditional telephony and enabled mobile telephony to serve a variety of needs in the marketplace, from high end corporate customers to mass market consumes. When mobile came to Africa, prepaid was the only way to go. It offers a number of important advantages for consumers including,

1. Ease of purchase, no start up fees or contracts.
2. No credit checks or bank references (In a country where few people have bank accounts or credit cards, it would not make sense to have this requirement).
3. Control of expenditure (Especially for people with limited income, it is important to be able to control spending).
4. Caller pays termination regime in African countries (One can receive calls even if no time is allotted. this means there are little to no barriers to be accessible to friends, family, and business/social contacts).
5. Customers can buy products with no commitments. There are no recurring fees.

For mobile operators prepaid has the following advantages

1. Prepaid customers have lower sales acquisition costs.
2. An increasing number of people in Africa obtain bank and credit card, allowing easy and inexpensive top up, making prepaid services more profitable.
3. Administrative costs are lower with prepaid. There is no billing or remarketing function.
4. There is no credit risk. Traffic is delivered only once the revenue has been received. For many postpaid operators, they consider developing a prepaid business because they can have a customer segment with relatively low risk.
5. Prepaid products generally don't have number portability requirements, allowing for lower costs. However some prepaid products have number portability.
6. Prepaid also offers a financial advantage from an operating expenditure perspective. It is the opposite with postpaid contracts where traffic is delivered first, and the payment comes later.

However prepaid also has certain disadvantages, and there are reasons why some mobile services are offered on a post-paid basis.

1. Prepaid customers have low ARPU (average revenue per user). Postpaid customers tend to spend more money than prepaid. However some prepaid customers can be cultivated to increase ARPU, and this can achieved through focused campaigns, discounts and new services. It requires a better understanding of customers.
2. Prepaid products are short-lived, and there is high churn as customers can easily switch to cheaper services. A post-paid contract allows the

operator to sign up a customer for a fixed period. It provides a guarantee of revenue.

3. Most prepaid products are SIM only. Users have to buy a phone separately. Post-paid contracts, as they come with a monthly fee, offer the facility to subsidize a handset. This is an important means to encourage users to take advantage of smartphones. Subsidies for phones can be difficult to optimize.
4. In practice it can be difficult to identify the users of prepaid products. Providers spend much energy on data mining. They understand their customer base in aggregate, but not by the user's identity. Should they want to target certain segments for specialized offers, they may have difficulty. For that reason many prepaid operators offer a bonus for self-identification, such as email addresses. At the same point, anonymity may be prized by some users.
5. The generally low barriers to entry in the prepaid market means there are more competitors, falling prices, and limited opportunities for profitability.
6. Prepaid customers generally have limited loyalty to their provider. Prepaid operators don't expend the effort to create loyalty programs. However versions of lite loyalty programs might be employed by some providers.
7. Prepaid products are subject to falling termination rates because regulators consistently lower interconnection and termination fees. Prepaid providers are more vulnerable to lower termination than post-paid providers. All things equal, prepaid providers have a diminished future of revenue given a set customer base. As such they have to evolve just to keep revenue constant.

As the examples have shown there are many aspects of postpaid that operators are trying to bring into prepaid, as a result there will likely be more hybrid products in future.

4.3 Mobile Financials: Africa vs. the Rest of World

Bank of America Merrill Lynch produces a report for investors of the telecom industry. They collect the financial information from the operators and aggregate it into a quarterly report. Data is only available for Algeria, Egypt, Morocco, Nigeria, and South Africa. These five countries make up 35% of Africa's population and illustrate the challenges across the continent.

The African subset is compared against figures for Europe (a subset of 15 countries), North America (US and Canada only), and Emerging Europe Middle East and Africa (EEMEA).

The key measures for the 5 countries in Africa include their high prepaid percentage, the high percentage of mobile revenue versus the countries' GDP, the high monthly churn, and the low ARPU. Prepaid is particularly high in Africa among other development regions; the corresponding figures for emerging Asia is 83% and Latin America as 80%. While post-paid services require some kind of identity card or bank reference, prepaid can generally be obtained just for cash. It can be surmised that prepaid is high because of the lack of financial and legal infrastructure, namely banks, regulatory authorities and so on. Not requiring a bank or ID cards lowers barriers, so many entrepreneurs start prepaid services.

That mobile services revenues makes up 2–3 times the percentage of GDP compared to other regions may be interpreted in two ways. On the one side it may be an indicator or relatively low economic development. On the other, it is an example that mobile itself is fueling the economic growth. Studies from the World Bank. (World Bank, 2012) and the UN. (United Nations, 2013) among other global organizations have reiterated this.

In general there are many competitors, low prices, and high churn in Africa. Compared to other regions of the world, this is indicative of an immature mobile market and suggests need for consolidation. The 9 mobile networks in Nigeria are not necessarily optimal. Nigeria is a country with low ARPU, so any one operator will have high costs to deliver traffic. Two operators joining together can improve the business case dramatically and so on. A country of Nigeria's size and population can be adequately served by 3–4 networks while preserving competition, so the additional revenue spent to build networks could have had other productive use in society.

A country tends to grow in mobile operators because issuing mobile licenses is a revenue strategy for the regulators and governments of many countries. While there can be legitimate reasons, regulators may prioritize short term income over efficiency.

When a market reaches 100% penetration, it is saturated. This means that people buy multiple SIM cards to perform mobile arbitrage. In many countries, operators offer plans with low prices or flat rate for on net traffic, and higher prices for off net traffic due to termination fees. This means that it costs nothing to call others who have the same mobile provider, but the charge only applies when calling others who have a different mobile provider. In practice people in many developing countries have more than one SIM card, and they switch cards based upon the people they call. Indeed some have business cards with multiple numbers for the various network providers. It is only reasonable that a consumer would take up such a practice when there are competitive offers in play, but it also shows that operators fail to get the full value of customers' mobile expenditures.

High churn rates can have a negative effect on operators in the short run, and for consumers, in the long run. To be sure, customers economize by switching operators based upon lower prices, and this is an understandable practice where incomes are low. The long term effect however is that few, if any, operators, will ever earn the margins necessary to invest in next generation infrastructure. If the market is characterized by a constant price war and no equilibrium, this "race to the bottom" will also hurt consumers for lack of investment. As such, the situation increased the likelihood that operators from outside Africa will purchase local players and provide more stability and investment.

Informa's World Cellular Data Metrics database provides another perspective. Just as operators experience in other parts of the world, the shift to data does not mean that operators will earn the same amounts of revenue for a given package. In fact the profitability of data is frequently less than voice. Data as a percentage of service revenue in Africa is about one third of the level in the Europe and North America. Naturally its growth will depend on the rate of deploying of mobile broadband infrastructure. In any event, just as operators in other parts of the world experience, voice gives a higher, more profitable return than data, and voice ARPU is on the decline.

The information shows a variety of important trends across Africa. Revenue from data and data as a percentage of total revenue is increasing. Meanwhile voice behaves as a counter trend. SMS revenue is holding constant, but this is likely because of the relatively low penetration of smartphones and broadband networks. The value of SMS in Africa should not be understated. In many places, SMS functions as de facto contracts. A fisherman bringing fish to market can ensure fair price in advance and buyers most keep the price in the message.

As broadband penetration increases, people tend to upgrade to a smartphone. Then they have the option to use an OTT service for SMS. Going forward, operators will likely price voice and SMS together to stave off the shift to a competitive service. The data also shows a slow but steady increase in mobile broadband subscriptions.

Table 4.2 Mobile statistics for selected African Countries

		Pop	Mobile Penet	Smart phone		Subscribers				Service Revenue		ARPU
	GDP/cap ($)	(mn)	%	%		(mn)	Prepaid	(US$ bn)	% of GDP	(US$)	Monthly Churn	No. of Players
Algeria	5413	36	100	5		37	93	3,7	1,9	8,54	2,3	3
Egypt	3072	82	125	10		103	97	5	2	4,07	2,9	3
Morocco	2956	33	118	6		38	95	2,9	3	6,36	1,9	3
Nigeria	1660	165	65	6		107	100	9,1	3,3	7,26	3,5	9
South Africa	7580	51	130	20		67	82	11,4	2,9	15,06	2,6	3
	4136,2	367	107,6	9,4		352	93,4	32,1	2,62	8,258	2,64	4,2
	AVE	TOT	AVE	AVE		TOT	AVE	TOT	AVE	AVE	AVE	AVE
Europe	33813	406	130%	39%		528,7	52%	165,2	1,2%	25,29	2,2%	
North America	48530	346	104%	39%		361,3	28%	196	1,2%	51,67	1,7%	
EEMEA	6884	719	110%	12		789,8	87%	89,3	1,80%	11,33	3,20%	

Source: Bank of America, Merrill Lynch 2013

The following diagrams illustrate some of the trends for Africa.

Metrics for Africa

Region KPI	Unit	Q2 2011	Q3 2011	Q4 2011	Q1 2012	Q2 2012	Q3 2012	Q4 2012	Q1 2013
Data as % of Service Revenue	%	11,55%	11,57%	11,89%	12,55%	12,97%	13,92%	14,32%	15,48%
Voice as % of Service Revenue	%	88,45%	88,43%	88,11%	87,45%	87,03%	86,08%	85,68%	84,52%
Non-SMS data revenue as a % of total data revenue	%	47,91%	46,20%	48,54%	56,14%	53,53%	55,68%	54,69%	53,97%
SMS Data as % of Data Revenue	%	52,09%	53,80%	51,46%	43,86%	46,47%	44,32%	45,31%	46,03%
Revenue - Data	USD m	1.716,58	1.724,00	1.787,55	1.868,94	1.948,19	2.108,95	2.211,12	2.390,33
Revenue - Voice	USD m	12.965,68	13.310,93	13.099,95	13.103,67	13.110,43	12.851,68	13.151,05	12.489,86
Revenue - Data (Non-SMS)	USD m	822,42	796,50	867,59	1.049,30	1.042,94	1.174,17	1.209,36	1.290,03
Revenue - Data (SMS)	USD m	894,16	927,50	919,96	819,64	905,26	934,78	1.001,75	1.100,31
ARPU - Data	USD	0,87	0,92	0,91	0,93	0,95	1,03	1,04	1,06
ARPU - Voice	USD	7,50	7,28	6,89	6,76	6,48	6,15	6,01	5,64
ARPU - Data (Non-SMS)	USD	0,38	0,40	0,45	0,47	0,46	0,53	0,58	0,63
ARPU - Data (SMS)	USD	0,49	0,51	0,46	0,46	0,48	0,50	0,46	0,43
SMS Traffic	Messages (millions)	25.811,52	27.602,55	27.947,68	28.288,49	29.441,83	30.009,65	36.074,64	36.398,33
MMS Traffic	Messages (millions)	836,87	874,93	903,19	904,16	907,19	942,75	931,53	937,53

Mobile broadband subscriptions	Unit	Q2 2011	Q3 2011	Q4 2011	Q1 2012	Q2 2012	Q3 2012	Q4 2012	Q1 2013
CDMA2000 1xEV-DO	Subscriptions	2.089.243	2.227.661	2.235.714	2.247.873	2.205.692	2.234.007	2.324.734	3.645.649
W-CDMA	Subscriptions	32.297.987	35.938.445	39.684.744	43.560.124	48.412.592	53.564.228	59.695.529	64.267.435
LTE	Subscriptions	0	0	0	0	2.790	6.195	25.468	229.480
Total	**Subscriptions**	**34.387.230**	**38.166.106**	**41.920.458**	**45.807.997**	**50.621.074**	**55.804.430**	**62.045.731**	**68.142.564**

Mobile broadband as % of total subscriptions		Q2 2011	Q3 2011	Q4 2011	Q1 2012	Q2 2012	Q3 2012	Q4 2012	Q1 2013
Mobile broadband %	%	5,8%	6,2%	6,5%	6,9%	7,3%	7,8%	8,3%	9,0%

Figure 4.2 Selected mobile metrics for Africa

Source: Informa, 2013

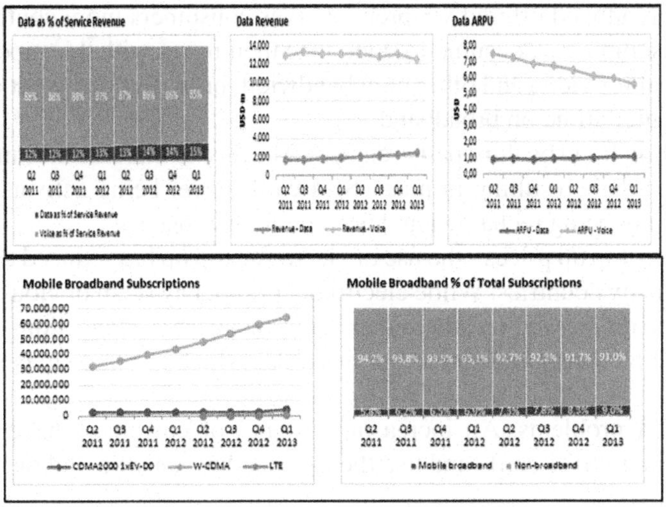

Figure 4.3 Selected financial metrics for African mobile market

4.4 Challenges for the Prepaid Market

There are a number of challenges for prepaid providers.

1. Heavy competition in the prepaid market lowers the prices for traffic. Operators offer more minutes and message at a lower price. It is good news for consumers, but it will tend to drive have an effect where the strong will become strong and the weak, weaker. Such environments are conducive to consolidation.
2. Regulators continue to lower termination and interconnection rates. This reduces revenue for prepaid providers.
3. Competition makes on net traffic cheap or free, again reducing operator's revenue.
4. A number player's offer cheap international traffic, including VoIP services such as Skype and MNVOs specializing in international traffic. These services compete with established operators.
5. Heavy competition creates bundling. Voice and SMS are bundled into low cost packages.
6. Heavy competition also creates churn. An operator with 10 million subscribers and a monthly churn rate will of 2.5% will lose 2.5 million customers a year. In practice an operator needs to gain 2.5 million customers per year just to keep the customer base constant.

7. Mobile broadband offered via prepaid gives consumers more control of their consumption. Consumers economize their usage with this model. It is not ideal for video and other high bandwidth providers that want users to spend more time on broadband.
8. Distribution fraud by third party retailers. The race to increase the number of SIM cards is resulting in many SIM cards never being activated, or alternatively being recharged and used more than once.
9. Decreasing top up prices - the use of electronic top up has reduced top up costs in many countries. While electronic top-up may seem obvious, it is difficult to do in a region where there are few internet connections. As a result, people must do their top ups at kiosks, which is a more expensive method.
10. Multi SIM problems. An increasing number of customers have two, three or more SIM cards and use them depending on who and/or when they call.

4.5 Marketing and Product Development of Prepaid

There is a growing market for both children and the elderly to use phones. Given their elastic demand, prepaid is a good solution. Safety, location-bases services and fixed budgets are important for these two segments. It may be the case the operators have to offer both services, prepaid and post-paid. For example postpaid customers may buy prepaid products, such as parents buying mobile services for children, or employers buying mobile services for employees. Both groups want a cap on expenditures.

There is a lot of innovation in both feature phones and smartphones. Prepaid operators can provide SIM cards with near field communication (NFC) functionality; these are debit cards that integrate with the phone. Email is another bundled service. Feature phones can be developed with the major operating systems (Android, Symbian, Windows Mobile etc) giving increased functionality. The phones have app stores with offer location-based and other services. These kinds of services are applicable on high end feature phones. Meanwhile low cost smartphones are entering the market.

Post-paid providers may look to prepaid offerings to grow revenue. There can be a variety of reasons. People may want more anonymity in their communications.

Operators also need to consider what mix of channels, whether operator owned shops, independent dealers, fast moving consumer goods dealers, or online web shops, when distributing their prepaid products.

4.6 Similarities between Europe and Africa

Europe and Africa have a number of similarities as continents, namely a wide diversity in geography and languages as well as a proliferation of small independent states. While the mobile market in Europe is more developed, it offers a number of lessons for Africa.

Europe is a continent with some 50 nations. While this makeup may be conducive to preserving cultural and political interests with sovereign states and distinct languages, it is not ideal for the formation of an efficient mobile market nor a digital single market. It can be contrasted to the United States, a federal country with 50 provinces under the jurisdiction of a strong central government. The US today accounts just 4% of the world's population, but enjoys 25% of the world's broadband investment. (Infonetics, 2013). There are essentially 4 leading mobile operators, 2 of which provide near 100% coverage across the country. A single country such as the US illustrates the benefit of economies of scale when building and running mobile networks. The US has the most extensive next generation mobile network, the highest penetration of mobile broadband subscriptions, and a high penetration of smartphones. This development is enabled by the fact that few mobile operators are able to make large spectrum footprints across the country which they can deploy their networks. Additionally the spectrum is offered on a technology neutral basis, so operators can upgrade with ease as well as obtain a license for 30 years, creating an incentive to steward the resource. Operators can also swap and trade spectrum. The key benefit for consumers is a seamless experience. No American pays roaming charges inside the USA, unlike the high charges consumers experience when the travel across Europe.

In contrast to the US, a continent with 50 or more countries may have challenge to create a large spectrum footprint. Offering spectrum in small pieces, on a short term basis, or on a defined technology may enable short term revenue for a telecom authority or government, but it does not serve a long term goal wide coverage or advanced technology. Additionally Europe's policy to manage competition through mandated access and structured entry may have enabled many players, but it does not help the continent achieve a high level of investment.

A legacy of history, each European country has its own telecom regulator, requiring that operators get new certificates for each and every nation they cover. Consolidation across countries is restricted. In practice this means that there are at least 50 individual copper networks, at least 50 mobile networks, and so on. Europe's telecom revenue and investment have been on the decline since the financial crisis. As Neelie Kroes, Vice President of the EU suggests, "Telecoms is no place for borders.... The missing cornerstone in this digital ecosystem is a Telecoms Single Market, which would permanently increase GDP by nearly 1 percent (Ł110 billion) if delivered." As such the EU has stewarded an effort for a digital single market. (EU Commission, 2013).

4.7 Competition, Consolidation, and Cooperation

It is a mantra of economics that competition is good. It is seen as the solution to many problems. However creating the conditions for competition is not so easy, and not all industries have the same features and characteristics. The elementary idea of a competitive market is one with many buyers, many sellers, perfect information, a uniform good, no taxation, and no barriers to entry. These conditions exist almost nowhere in the world, but the example of farmers selling crops in a marketplace is often used to illustrate the idea. Within a given area, two or more farmers may sell the same crop. They come to a marketplace and the interplay between buyers and sellers creates the price. A third farmer or a new commodity changes the mix in the marketplace, as does the different decision of the farmers. For example, one farmer may use a different technology to produce the crop more efficiently, and therefore offer it at lower price for example, and so on.

Telecommunications is very different from farming. It is a highly sophisticated service requiring many inputs and technology. Additionally it needs a substantial economies of scale in order to achieve efficiency. For example, a square kilometer of land will likely not be a large enough area to justify creating a single telecommunications operation. To justify the investment, a large area is generally required.

On account of the high fixed costs and entrance barriers, telecommunications has traditional been run as a monopoly. It was not practical to have multiple firms erect their own telephone poles and wires, so the right is typically offered to one firm in exchange for certain covenants to serve a defined area or population and specific rates. However the rise of new networks, including mobile communications, has challenged that traditional model to a certain extent.

Mobile networks require spectrum, infrastructure, terminals and subscriptions. While under a traditional telecom model these inputs would be offered under a vertically integrated model, mobile networks are highly diversified with a number of competing firms at each level. To begin, the finite resource of spectrum must be allocated. A number of firms may compete in an auction for different bands, or in a less efficient way, the government may award spectrum based upon application. A mobile operator will typically contract with an infrastructure provider to build mobile infrastructure including masts and towers to provide coverage for the given area. Makers of mobile terminals will offer equipment to end users either directly or through a resell agreement with operators. End user subscriptions can be offered directly to consumers or through wholesale agreement with mobile network virtual operators (MVNOs). As for services, they can be provided by either network or virtual operators, or they can be offered by third party in so-call "over the top" (OTT) or pass through services.

While the challenge is not limited to Africa, it can be the case that for historical reasons a variety of operators have different pieces of spectrum, which may or may not comprise a substantial footprint. It may be that equipment to provide 2G mobile service is deployed, and there is a market demand to upgrade to 3G or 4G. It may then be the case that allowing one operator to purchase another and combine the two sets of spectrum can allow for a more efficient deployment of the next generation technology. This acquisition would be further strengthened by the fact that the acquiring operator could increase its chances for financing by having a large customer base, and theoretically a larger projected revenue. In practice, the small operator with limited spectrum may have difficulty asking for a loan to upgrade to the next generation technology. It may not be feasible without substantially raising prices or agreeing to unusually high interest rate, which might also increase prices. The model is further complicated by the fact that a larger operator is likely to be able to negotiate better volume discount for next generation equipment. Additionally should the next generation equipment be associated with offering advanced services, it is usually the case that a larger provider will be able to make larger outlay in the service platform. These are just simple examples, but they illustrate why consolidation is frequently desirable in an economies of scale market with high fixed costs and high barriers to entry.

It is not exactly the case that a straight M&A model will work for African operators. Operators could cooperate to share infrastructure. Having two operators use the same mobile mast is an increasingly common practice. This could also work for a backbone/backhaul. Two operators could join production units for different countries for example. However many forms of cooperation can also lead to consolidation.

4.8 The Case for Consolidation

Consolidation is a widely discussed topic in business and strategy literature, but it has a certain nuances and implications for the mobile industry. Just to review, the arguments for consolidation for why one company would by another include gaining more market share, deploying better business models across a larger customer base, accessing new technologies, getting better terms for financing, finding hidden or nonperforming assets belonging to a target company, and winning better bargaining power over their suppliers and clients. Another key benefit of consolidation is lowering, or making more efficient, administration costs, for example, the same financial department can serve two companies.

When applied to network industries such as telecommunication, consolidation can have some important benefits including reduced operating expenditures on network operators, reduced capital expenditure with fewer sites (or the removal of redundant sites), reduced marketing costs (fewer brands to market), and better utilization of spectrum and infrastructure investment. In mature economies, sales and marketing costs can consume up to 25% of an operator's revenue, so reducing this line item through a consolidation is an attractive proposition.

It should be noted that consolidation appears in varying degrees, from acquisition to network sharing (joint venture). Management consultants Arthur D. Little describe some opportunities and challenges in consolidation. One report *Network Cooperating: Making It Work and Creating Value* describes network sharing as the "practice of Mobile Network Operator (MNO) sharing part of its Radio Access Network (RAN) with another MNO. Often the scope of network cooperation ranges from passive RAN cooperation—the sharing of sites, tower structures, shelters, power and cooling—to active RAN cooperation—the additional sharing of backhaul transmission, backhaul fiber, antenna, and site electronics". (International Telecommunications Union, 2013).

In the context of a mobile network, consolidation and joint ventures would allow savings on operating expenditure (land rental, electricity etc), manpower to maintain sites, and saving on duplicate infrastructure. In addition the consolidation can be enriched by selling decommissioned sites for cash. (International Telecommunications Union, 2013). The company explains, "When two RAN networks are combined, the geographical size and scope of the resulting network is greater than each individual network. Each MNO now gains access to a geographically larger network and the complementary strengths of the other, such as access to new sites/towers, spare backhaul capacity etc.". (International Telecommunications Union, 2013).

4.9 Why not have the Government Provide Mobile Communications?

There is a perspective that telecommunications is an essential facility and that operators should be responsible to the community or citizens, not shareholders. While there may be merits to this thinking, it should be noted that most countries have moved away from government-led telecommunications models. While a few remain, they are the minority.

Africa does not have the same tradition for the state-run monopoly. There were few instances of a large, administrative government in the first place. A number of African countries have been characterized by lack of stable governments, making a state run mobile operator even less likely. The proliferation of successful mobile models in many countries suggest that the market led mobile services is the better way to go for Africa. In any event, mobile penetration is high in Africa, so there is evidence that the market can work. As for problems with too many networks, it may be the issue of regulatory capture, rather than the market itself.

4.10 Carrier's Carrier Model

Given the challenges in Africa for build and operate mobile infrastructure, it may be helpful to investigate the carrier's carrier model of mobile infrastructure. A government might consider this model to facilitate the rollout of a next generation technology or as a means to serve an area that has geographic or financial challenges. In the carrier's carrier model one operator alone is dedicated to the building of mobile infrastructure. It then resells access to the infrastructure to other players, but does not participate in offering services. It only offer infrastructure. With improving technology the model is increasingly feasible. A single base station can accommodate a variety of mobile technologies (for example GSM, UMTS, LTE) and frequencies (for example 800, 900, 1800, 2100, 2600 MHz). The model is already underway in Mexico, and is seen as an important way to create competition where the incumbent provider has 70% of the market.

4.11 Investment and Corporate Structure

Investing in mobile in Africa is attractive, and the complexion of operator groups is not unlike the rest of the world. There are some African based operators as well as European, Middle Eastern, and Indian. The telecom research firm Infonetics provides a comprehensive global review of capital expenditure in the mobile industry, but African investments are included under the umbrella of international operators. It is difficult to estimate the exact amount for Africa but it is likely to be in the low billions of dollars. It is important to note that while Westerners may associate CAPEX with network upgrades in 4G or fiber to the home, many investments in Africa are to upgrade more modest technology, from 2G to 3G for example. CAPEX also goes in cycles. An operator in an African country many spend a high amount for 2–3 years, say tens of millions of dollars, then reduce that amount to a few million per year.

The following chart from World Cellular Investors shows the largest operating groups in Africa by subscribers. These are also the groups that will typically invest in the region.

South Africa is the home to the headquarters of the three largest network operators in Africa. They include MTN, Vodacom, and Telecom. MTN is the largest African-based mobile operator. It has over 100 million subscribers and operates across 21 African countries, including South Africa, Nigeria, Ghana, Sudan and Congo. British Vodafone, the world's second largest operator, operators in Ghana, Egypt, and South Africa. It has a majority stake in South Africa's Vodacom, and has operations in Tanzania, the Democratic Republic of Congo, Mozambique and Lesotho. (Thomas, 2013) Telecom, the fixed lined incumbent, is moving into mobile. It is 39% owned by the South African government. (IT News Africa, 2012).

Safaricom is the largest mobile service provider in Kenya with 12 million subscribers. It is a daughter company of Telkom Kenya. Vodafone has a 40% share in the company. Safaricom was the first east African nation to provide its users with 3G mobile networks. (IT News Africa, 2012).

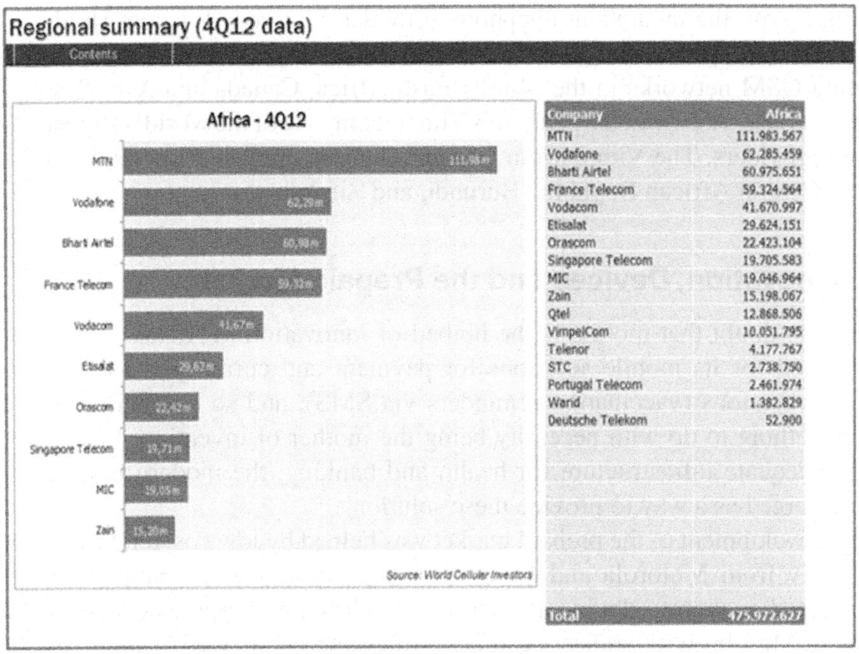

Figure 4.4 Subscriber data for selected Africa mobile operators

Source: (Informa Telecoms & Media, 2013)

For European operators, not only is there geographic proximity and shared time zones as well as language, it offers a ways to offset losses from Europe. Orange, formerly France Telecom, is the fourth largest operator in Africa, and the area is its fastest growth region, on track to a doubling of revenue to Ł7bn by 2015. It aims to increase revenues from the countries Tunisia, Ivory Coast, Guinea, Cameroon, Madagascar, Togo, Mauritania, Burkina Faso and Gabon.

Middle Eastern and Indian operators also have an affinity for the region. The Indian operator Bharti Airtel, the third largest operator in Africa, spent $10 billion to acquire a stake in the region in 2010. (Thomas, 2013) Etisalat, the Emirates-based telecoms group bought a stake in Maroc Telecom, as did the French Vivendi. Maroc Telecom is a large provider that also works in Burkina Faso, Gabon and Mali.

Orascom Telecom, based in Egypt, operates GSM networks in Algeria (Djezzy), Egypt (Mobinil), Pakistan (Mobilink), Bangladesh (Banglalink), Tunisia (Tunisiana), and Zimbabwe (Telecel Zimbabwe). Also in Egypt is

Telecom Egypt, the incumbent telephony provider. Orascom Telecom Holding, which is the parent company of Orascom Telecom, is one of the largest operating GSM networks in the Middle East, Africa, Canada and Asia. The company is also 51% owned by Russia's Vimpelcom, one of the world's largest telecom providers. The VimpelCom Group includes companies operating in Algeria, Central African Republic, Burundi, and Zimbabwe.

4.12 Innovation, Devices and the Prepaid Market

There is no doubt that mobile is the hotbed of innovation in Africa. Africa is renowned for its mobile solutions for payment and currency (M-Pesa), health applications (vaccination reminders via SMS), and so on. This may have something to do with necessity being the mother of invention. Indeed lacking adequate infrastructure for health and banking, the modern mobile phone emerged as a way to provide these solutions.

The development of the prepaid market was helped by low cost terminals, particularly from Motorola and Nokia. Many of these feature phones are sophisticated enough to double as smartphones. They come preloaded with the Opera Mini browser so users can access the internet as well as apps for Facebook, Skype and WhatsApp. Opera's browser is so flexible that it can compress data between a feature phone and 2G network enabling the users to visit the web.

Low cost smartphones with Android operating systems have started to enter the market. These phones manufactured in China, Taiwan, and India, are helping speed the adoption of mobile broadband. Africa also has a robust market for used phones. When one upgrades a phone, the phone is not necessarily thrown away. It may be placed on the secondary market or used by a friend or family member.

M-Pesa (m for mobile and pesa for money in Swahili) emerged from the prepaid market in Kenya where users transferred airtime as a means of currency in a situation where banks and bank services were virtually non-existing – especially in the rural districts. Today M-Pesa accounts for a third of Kenya's GDP. Users don't need a bank account to transfer money, just a mobile device and subscription as well as a passport or national ID card. M-Pesa is a service run by two Africa operators, Safaricom and Vodacom, and is available in Kenya, Tanzania, Afghanistan, South Africa, and India. Financial services include deposits, withdraws, bill payment, and in some marking, an interface with a banking institution. M-Pesa is not just a currency, it is a form of microfinance. Users can start a business with an advance for the equivalent of $100.

Another innovation allows African to access Facebook with an ordinary mobile phone and no internet connection at all. Operators serving the 15–24 aged segment now have a novel, revenue-generating solution. U2opia Mobile created a USSD platform which allows users to connect to Facebook with a basic mobile phone and no internet connection. Its application Fonetwish (http://www.fonetwish.com) allows users to make status updates, browse news feeds, post on walls, send/accept friend requests, send messages and view notification on Facebook.

USSD (Unstructured Supplementary Service Data) is a low bandwidth data service that allows information to be shared on something as basic as a 2G network with no Internet connectivity. It feels like engaging with Facebook via SMS, and it works on any phone. The suite of U2opia Mobile applications includes solutions to access Google and Twitter without a data connection.

Competition in Africa for mobile subscriptions is fierce, so the Fonetwish application is an important churn reducing enhancement. Given that internet penetration in Africa is low, the opportunity for customers to upgrade to a smartphone with a data package is limited. Operators with Fonetwish, not only keep their customers, but they may recover marginal revenue that might have gone to internet cafes instead.

Naturally the bare bones service does not support video or images, but that is not a deterrent for this target audience of Fonetwish. Indeed their expectation from Facebook, Gmail, and Twitter is not the same as a typical Westerner with a smartphone. So the USSD solution is an upgrade over no Facebook access at all.

While the majority of Facebook users in Western countries are middle–aged, Facebook users in Africa are primarily teenagers and young adults. Naturally this is an important segment for operators who want to establish their brand with consumers as early as possible.

4.13 Solutions

Most African mobile operators can reduce the size of their prepaid sales and marketing costs and reduce costs on servicing customers for topping-up or recharging their mobile accounts. Just within the topping-up area, cost reductions could be 2–20%. Most operators battle against high churn figures. A monthly churn of over 3% is normal for prepaid customers. There are a number of tools that can help reduce an operator's monthly churn by between

5–10 percentage points, provided that the mobile operator understands how to use the tools.

Together cost and churn reductions can improve overall EBITDA 2-3%.

Following are discussions for how an operator can improve its prepaid business case and reduce the negative impacts of customers using multiple SIM products.

4.13.1 Ways for an Operator to Improve the Prepaid Business Case

A prepaid provider can improve its business case in a number of ways. Here are some of the possibilities.

1. Bundle prepaid plans with special features such as extra minutes, SMS, free/low cost international traffic etc. Bonuses on recharge are prevalent across Africa, particularly from challenger operators. These increasingly include data, and promote top-up via mobile money. In Nigeria, Bharti is rewarding data top-ups with lower voice pricing. Also we note a trend towards bundling of voice/SMS/data in prepaid markets.
2. Incentivize the customer to top up on the price plan they want
3. Make available various channels for top up and incentivize customer to use the least expensive methods. Following are the various methods and the operating costs. The other problem with scratch cards is that they are most expensive of the prepaid products and they are an effective tax on people who don't have banking or internet facilities.

    Scratch cards through retailers (20–25%)
    Electronic top up through retailers (processed by cash register or terminal)
    Electronic vouchers
    SIM solution on phone (8–12%)
    Automatic teller machine (ATM) (2–5%)
    Credit/debit card or bank payment (1–4%)
    Bank's website
    Peer to peer (1%)
    Operator's website (less than 1%)
    Fixed invoice (less than 1%)

The following example shows how an operator can hold revenue constant but increase profitability by lowering top up cost by 40%. (Strand Consult, 2012).

Table 4.3 Summary of mobile top up costs for sample operator

Type of Top Up	Cost of Revenue	Share of Top Up Types	Old Revenue (USD $)	Old Top up Cost	New share of Top Up type	New Revenue (USD $)	New Top up Cost (USD $)
Scratch cards	25%	20%	2000	500	0%	0	0
Elektronic	10%	67%	6700	670	55%	5500	550
ATM	5%	10%	1000	50	15%	1500	75
Credit card	3%	3%	300	9	25%	2500	75
Peer to peer	1%	0%	0	0	5%	500	5
Total		100%	10.000	1229	100%	10.000	705

Source: Strand Consult

4. Operators can create loyalty programs Loyalty program can work in a variety of way. A promotion can include free airtime. A reward for loyalty can be free or reduced airtime.
5. Operators can offer packages that allow changing plan at the time of top up, balance transfers to different plans, and remarketing to prepaid customers from previously collected emails.
6. The operator can make its product stickier. This could take a number of forms. Operators can offer packages for families and groups of friends. This increases the number of customers and decreases churn. The operator can focus on getting customers to increase the apps on the phone; this makes it less likely for a user to switch. The customer can also offer an API facility providing billing and distribution for various apps. As the users accesses these valuable apps on the phone, there is lessened chance for churn.
7. Make partnerships with banks. Financial institutions have capital and credibility. Naturally banks can also offer their own mobile solutions. The prepaid operator would need to provide a compelling reason to cooperate.
8. Make partnerships and discounts with retails chains to lower distribution costs. Retailers are generally available where customers are, though they tend to charge high commissions for the privilege of distribution.

4.13.2 How to Address the Multi SIM World

From a consumer perspective, there are good reasons to have many SIM cards, most important is to save money. In stark contrast to subscribers on post-paid contracts, the savings are not worth the cost of time. Most post-paid subscribers just have one SIM card. There are additional reasons why Africans may have multiple SIM cards. They include taking advantage of different benefits from many operators in the marketplace, having separate phones for personal and professional domains, to ensure mobility in places where there is not consistent coverage by the same operator, and to remain anonymous on the mobile phone.

From an operator's perspective the multi SIM world is detrimental to revenue, and it should attempt to gather as much traffic on its own network and reduce the number of outside SIM cards. Some ways to do that include

1. Enabling many service on the SIM card, minimizing customers' use of competitor's products.
2. Enabling services in such a way that it encourages keeping the SIM active and having the phone on, such as family messaging and email.
3. Providing an email service will forcing the SIM to be active to receive e-mails, encourage data usage, and restrict the service such that it cannot be ported if the customer leaves. Another attractive feature, a push email service send the user's mail to the phone.
4. Location-based information services are helpful for parents who want to stay up to date on children's locations and other parties who want to track each other such as friends, spouses, and employers/employees. They force the SIM to be active and encourage the user to keep the same provider. Such services can also be valuable safety precautions for people who have to travel long distances. They can be located by emergency professionals.
5. Prepaid mobile payment solutions are tied to the balance on the SIM card. They are good solutions for children and employees where a fixed budget is important. The SIM card needs to stay active in order for the payments to work.
6. Operators can also offer a service to integrate old phone numbers on the same SIM so the user can reduce the number of phones.
7. An operator can offer a phonebook solution where numbers dialled are automatically registered. This increased the probability of incoming calls.
8. The operator can also enable a receiver pays SIM. This is helpful for children and family members to ensure that calls are made even when there are no more credits left. It could also enable a text service in which the receiver pays. The default message could be "call me free".
9. The operator can offer a variety of member services such as bonus to activate a dead SIM, lower rates for bringing in new people into the network.
10. When onboarding a new customer, an operator can offer a transition service where called numbers are automatically ported to the new SIM.

4.14 Conclusion

Africa is a continent of dynamic mobile activity. The prepaid market is the prevailing market to provide mobile subscriptions. This paper offers a number of suggestions to improve the mobile market for the benefit of consumers including consolidation of operators to improve the rate of investment, reducing the cost of top up, and reducing churn. There is no doubt that Africa has the demand for mobile broadband, but the mobile industry will need to become more efficient to deliver the broadband and mobile products in the future.

4.15 Appendix

The following data come from the ITU website, under the link labelled Percentage of Individuals using the Internet http://www.itu.int/en/ITU-D/Statistics/Documents/statistics/2013/Individuals_Internet_2000-2012.xls

	Percentage of Individuals using the Internet												
	2000	2001	2002	2003	2004	2005	2006	2007	2008	2009	2010	2011	2012
Algeria	0,69	1,37	2,37	3,35	11,61	15,08	19,77	21,50	33,10	41,30	52,00	53,00	55,00
Angola	7.40	11,02	14,30	14,59	24,27	25,41	34,95	38,38	40,44		41,00	43,16	47,08
Benin	0,64	0,84	2,72	4,04	11,92	12,75	13,66	16,03	18,01	25,69	31,42	39,83	44,07
Botswana	2,75	4,30	5,25	6,49	8,53	9,66	12,99	17,10	27,53	34,07	36,80	39,10	41,44
Burkina Faso	7,28	8,78	10,25	12,19	13,69	15,17	16,70	20,22	21,81	22,51	28,33	34,95	41,39
Burundi	5,35	6,35	6,71	7,01	8,43	7,49	7,61	8,07	8,43	10,00	24,00	33,97	41,00
Cameroon	2,20	3,62	5,35	5,98	6,59	6,87	11,04	16,30	23,20	24,80	26,53	34,00	38,93
Central African Rep.	0,06	0,09	0,32	0,56	1,29	3,55	5,55	6,77	15,86	20,00	24,00	28,43	32,88
Chad	0,32	0,62	1,21	2,94	3,02	3,10	7,53	7,95	8,67	10,04	14,00	28,00	32,10
Congo	0,03	0,14	0,44	0,54	0,79	1,29		8,66			16,70	19,00	21,00
Congo (Dem. Rep.)	0,93	1,28	1,82	2,44	3,23	3,70	3,70	4,10	6,85	8,94	11,04	18,13	20,78
Côte d'Ivoire	0,19	0,37	2,24	2,81	3,53	3,92	4,30	4,72	9,00	10,80	14,00	14,00	19,86
Djibouti	0,40	0,98	1,01	2,10	4,39	4,79	5,61	7,70	10,60	14,50	16,00	17,50	19,20
Egypt	0,15	0,20	0,83	1,19	1,72	1,83	2,72	3,85	4,27	5,44	7.80	14,11	17,11
Eritrea	0,40	0,80	3,99	6,39	6,56	8,02	9,79	10,85	11,40	11,36	11,50	15,70	17,09
Ethiopia	0,11	0,14	0,27	0,37	0,46	1,14	1,91	3,20	4,60	6,00	10,00	14,78	16,94
Gabon	0,49	0,65	1,59	2,20	4,63	5,84	7,38	9,45	10,18	11,23	12,50	14,00	15,23
Gambia	0,16	0,24	0,38	0,46	0,72	1,74	2,53	3,67	7,90	9,78	12,50	13,01	14,69
Ghana	0,19	0,23	0,48	0,98	2,01	2,85	4,16	4,87	5,55	6,31	10.00	11,50	13,47
Guinea	1,64	2,42	2,63	3,36	3,80	4,01	4,40	4,84	5,33	6.50	11,60	12,00	12,94
Guinea-Bissau	0.92	1,34	1,80	2,44	3,31	3.80	5,24	6,21	6,88	7,63	9,20	10,87	12,45
Kenya	2,90	3,43	3,39	3,35	3,30	3,26	4,29	5,28	6,25	6,15	6,00	8,00	11,50
Lesotho	1,22	1,35	1,94	2,66	2,98	4,89	5,49	5,77	6,21	6,70	7,23	8,00	8,62
Liberia	0,19	0,34	0,49	0,63	0,78	0,95	1,27	1,62	2,26	4,00	6,50	7,00	8,27

	Percentage of Individuals using the Internet												
	2000	2001	2002	2003	2004	2005	2006	2007	2008	2009	2010	2011	2012
Libya	0,06	0,24	0,29	0,36	0,43	0,56		2,12	4,50	7,70	8,00	7,00	8,02
Madagascar	0,03	0,03	0,16	0,46	1,08	1,46	2,01	2,76	4,29	4,50	5,00	5,60	6,11
Malawi	0,25	0,28	0,36	0,59	0,98	1,40	2,03	2,93	3,40	3,84	4,30	5,00	5,70
Maldives	0,11	0,16	0,26	0,42	0,68	0,85	0,84	0,91	1,56	2,68	4,17	4,30	4,85
Mali	0,21	0,26	1,08	1,53	2,18	2,58	2,98	3,45	3,58	3,72	3,86	4,22	4,59
Mauritius	0,13	0,16	0,22	0,28	0,35	0,38	0,43	0,97	0,70	1,07	2,26	3,33	4,35
Morocco	0,80	0,90	1,00	1,20	1,50	1,80	2,00	2,20	2,40	2,60	3,00	3,50	4,00
Mozambique	0,23	0,36	0,70	0,95	1,18	1,27	1,54	1,79	1,85	2,24	3,13	3,50	3,80
Namibia	0,02	0,03	0,03	0,03	0,03			0,55	0,53	0,51	2,30	3,00	3,79
Niger	0,08	0,16	0,20	0,37	0,40	0,47	0,63	0,75	0,92	1,13	2,40	3,00	3,73
Nigeria	0,05	0,08	0,13	0,15	0,22	0,27	0,31	0,38	1,00	1,80	2,00	2,20	3,00
Rwanda	0,23	0,30	1,02	1,35	1,81	1,90	2,06	2,21	2,35	2,30	2,45	2,67	2,89
Senegal	0,23	0,40	0,50	0,76	0,85	1,04	1,52	1,80	1,90	2,00	2,10	2,20	2,38
Seychelles	0,14	0,19	0,23	0,31	0,43	0,51	0,73	0,81	1,57	1,80	1,90	2,00	2,17
Sierra Leone	0,04	0,05	0,17	0,32	0,36	0,40	0,58	0,85	1,19	1,50	1,70	1,90	2,10
Somalia	0,20	0,22	0,34	0,42	0,53	0,57	0,61	0,65	1,65	1,63	1,70	1,90	2,05
South Africa	0,01	0,01	0,09	0,13	0,20	0,24	0,30	0,37	0,44	0,56	0,72	1,20	1,68
Sudan	0,10	0,18	0,40	0,45	0,51	0,54	0,64	0,78	0,92	0,94	1,00	1,30	1,49
Swaziland	0,02	0,04	0,07	0,11	0,16	0,22	0,31	0,37	0,45	0,54	0,75	1,10	1,48
Togo	0,04	0,11	0,13	0,16	0,19	0,22	0,29	0,39	0,70	0,76	0,83	1,30	1,41
Tunisia	0,02	0,08	0,12	0,38	1,05	1,08	1,10	1,12	1,14	1,16		1,25	1,38
Uganda	0,12	0,16	0,18	0,19	0,20	0,22	0,23	0,24	0,25	0,26	0,58	0,90	1,30
Zambia	0,08	0,11	0,12	0,20	0,35	0,54	0,66	0,70	0,81	0,90	1,00	1,11	1,22
Zimbabwe	0,14	0,16	0,23									0,70	0,80

5

Mobile Devices Vulnerabilities: Challenges to Mobile Development in Africa

Abdullahi Isah
CMI, Department of Electronic System
Aalborg University, Copenhagen
Denmark
abdullah@cmi.aau.dk

5.1 Introduction

Mobile phones are revolutionizing the way Africans conduct their daily affairs ranging from social relationships to financial transactions. Today, in most African countries, mobile phones are used as for financial transactions. Users can send money from one place to another via text messaging. Moreover, access to other market information has also been made possible with mobile devices. An example is the agricultural market information as most Africans are farmers that live in rural areas. Thus, due to the availability of internet enabled mobile devices, the African farmer can exchange and share market intelligence to know who is looking for what, where, when, and for how much. Unfortunately, most African mobile users under evaluate the threats of cyber-attacks on mobile devices. Social engineering attacks through SMS are rampant in the African society. The false SMS, in most cases, appear to come from well-known companies, mobile network operators, banks, or even friends and relatives. In reality, none of the aforementioned entities sent the messages. In most cases, the messages are accepted by gullible users. Moreover, software-based threats are also rampant on mobile devices leading to intrusions into the user's private data and stealing the user's sensitive information. Hackers easily eavesdrop on a user with the help of virtual base stations; or infect user's device with a virus which cannot easily be traced by antivirus. The viruses provide hackers with instant and complete control of the users' devices. The activities of the user are then monitored and used for the advantage of the hacker. More disastrous is the virtual base station that allows the owner to connect to the user's mobile phone and download information on the user's contacts, text messages, and other sensitive information. All the information stored on the user's phone can be downloaded by the hacker. Similarly, the hacker can upload all information on the user's mobile phone in few seconds. The security challenges on mobile devices in Africa are crucial to the development of communication in African society. The subsequent sections shall further explore the vulnerabilities and offer some solutions.

5.2 Mobile Devices and Vulnerabilities

The shift from PC based user interactions to mobile device user interactions has equally shifted the threat and security challenges to mobile devices. The increasing access of social media services via mobile devices makes mobile devices vulnerable to threats and attacks. Mobile devices are now targets of malware threats because they are not very secured and susceptible to end user manipulation through social engineering tactics. Africans are active on social media platforms such as Facebook, Twitter and YouTube. These platforms are sources of malwares. Social engineering on this platform occur through phishing, email, direct messages, friend requests, notification, wall posts, tweets, shortened links, and click-jacking.

In Africa there are five main areas of mobile device vulnerabilities and these are:

1. Stolen devices.
2. Rogue applications.
3. Critical information stored unencrypted, thereby given rise to stolen data,
4. Social engineering phishing,
5. Visit to malicious websites, and malware.

The history of twitter hacks started in 4/2007 with SMS updates vulnerable, 8/2008 Trojan download attacks begin; 2/2009 click-jacking attacks begin; 4/2009 XSS worm released; 4/2009 internal admin tool hack; 6/2009 Trending topic abuse begins; 7/2009 koobface; 1/2010 banned 370 passwords; 5/2010 force follow bug; 9/2010 Mouseover exploits found; 3/2011added an option that requires SSL; 9/2011script_kiddiez rampage. Facebook is highly distributed and in real-time the users of Facebook are more susceptible to social engineering. Facebook accounts are highly personal, that means they can disclose sensitive information that criminals find useful on targeted tasks. Facebook is faced with mass phishing spam. Facebook malware hacks include likejacking – waiting for users to click on them thereby bringing unwanted consequences. Other threats to mobile devices include Photo tagging, comment jacking, rogue application features; self-cross-site scripting. The most notorious social malware is koobface. It is a botnet that exploit users on social network through four stages of attack. In the first stage fake posts and comments are redirected, in the second stage malicious bit.ly and BlogSpot URLs redirects to the third stage where hijacked website with JavaScript will direct the user to the fourth stage where a server finally spreads koobface using different themes, Fake YouTube video tricking users to click on links.

In Africa, the use of mobile devices for personal, business and official functions is on the increase. Africa is a big market for mobile devices and the market share continues to grow with the increasing demand for smartphones and other mobile devices. Mobile devices offer numerous benefits as well as facilitates interactions among Africans. However the desirable benefits could not be realized without risking data and valuable information to 'superhighway robbers', otherwise known as hackers and attackers. The widespread of mobile devices in the African society, coupled with African governments' policies on digitalization of financial, academic and business activities, the situation turned out to be a harvesting ground for criminals using a social system of security known as social engineering to lure the African mobile user to give out information or act in the way the criminals wanted. While the African society is moving towards electronic of things, cyber criminals are sleepless to thwart the realization of this laudable goal. Cases of mobile devices threats are common wide spread and on the increase.

Mobile device malware started in 2004. Malware spread was on the hyper-connected mobile devices, primarily Android. The history of malware on mobile devices started with Symbian, J2ME iOS, BlackBerry, and WP. Mobile device malware are propagated by 'App Store' downloads, sending malicious links via SMS, email phishing, and browser vulnerabilities. In 2010, Juniper recorded 2.2% infection rate across all mobile Oss. 17% were SMS Trojans, 61% were spyware apps. History of Android Malware started in January 2010 with 09Droid, which was the first Android bank phishing application. In July 2010 there was Tapsnake, the first Adroid GPS spyware. In August 2010, Geinimi was born, as trojan that spread through pirated applications. In January 2011, there was ADRD, a search engine positioning malware. DroidDream came in March 2011 as first Android Market trojan. BigServ also arrived in March 2011 as pirated malware removal tool. In May 2011 Surge arrived in different forms like Zsone, Spacem, and LightDD. In June 2011, DroidKungfu came with its C&C controlled botnet. Basebridge also arrived in June 2011 with rage against the cage exploit. In June 2011 Plankton came with first jar downloader. GGTracker also came in June 2011 as first drive-by Trojan. In July 2011 SpyGold came with SMS and phone spyware. Also in July 2011 ZITMO came in Android in repackaged applications. Nikispy which steals GPS and Wi-Fi location, surfaced in July 2011. SMSsniffer arrived in July 2011 as stealer of mTANs. In September, 2011 SPITMO came in as spyEye variant.

In October 2011, JiFake arrived and posed as Mobile Messenger. In December 2011, RuFraud came in as Android Market Trojans. The history of iOS malware Hacks started in 7/2007 with First iPhone web vulnerability. In 2/2008, there was First iPhone software jainbreak. In 7/2009 malformed SMS vulnerability came in. In November 2009, Ikee was born as worm for JB iPhones. In the same 11/2009, Duh was also born for JB iPhones. In 7/2011 JailbreakMe arrived as website exploits, case example is Safari PDF vulnerability to JB iPhones. In 7/2011 Instalstock was noticed in AppStore applications, exploits code-signing vulnerability. In same July 2011, SSL Man-in-the-Middle bugged in. The Kaspersky lab collection reported that in 2012 40,059 of the 46,415 modifications and 138 of the 469 mobile families were added in their database

The following table displays the modifications and families of mobile malware in Kaspersky Lab's collection as at January 2013.

The growth of mobile threats is out weighing that of PC based threats; and with the African society turning into cashless and digital society, malicious applications that intercept SMS/text messages would compromise most of the authentication process of online financial transactions and online banking. The situation of mobile threats is not a healthy one to the growth, development and profitability of mobile communication in Africa.

Table 5.1 Mobile device vulnerability vectors

Platform	Modification	Family
Android	43600	255
J2ME	2257	64
Symbian	445	113
Windows Mobile	85	27
Others	28	10
Total	46415	469

Source: Kaspersky Lab, 2013.

5.3 Applications Vulnerabilities

Software runs the world. Today's mobile and smart phones cannot attract the high turnover without applications. Mobile applications are increasingly becoming popular, enticing and entertaining, and as well becoming a priority attack vector. An average mobile user has at least 75 applications on his/her device. Most of the applications are developed by third parties and malicious software may be embedded into them and when the user download them, the user is exposed to more threats and attacks. Security in terms of mobile devices is not yet well advanced, and users expect the ISP to provide some level of security, which is not always the case. Smartphones are many times more useful than any device now available for human use. Mobile device is no longer just phones, they are guides, assistance, advisers, and best friends. A compromised phone can leak information about the user. The criminal searches for victims from social and professional networks and send SMS messages enticing them to download patches and installs critical updates. Just with a few clicks, the user installs the spyware. The criminal then receives a text message that the malware has been successfully installed. The attacker has full control over the victim's phone instructs the phone to intercept every incoming calls and every incoming SMS. This is targeted persistent attack, which is common with mobile phones and is driven by applications downloads.

5.4 Twitter Security

Twitter is a communication platform for people to share information in a more concise and manageable messages of about 140 characters. People twit on what is going on in their social, personal and business life by sending short messages and blogs. Twitter has many enticing applications, and like any social media it is vulnerable to applications and user exploits. For example, in February 2013, more than 200,000 twitter accounts were compromised through Java 7 exploits. The accounts of the victims were remotely controlled without authentication. The affected OS were Windows, OS x, and Linux. Since the source of the attack was through java 7, users can disable it. In cases where it is required before some services can be accessed, the update 11 should be installed. Alternatively, the applets can be set not to be accessible. Similarly, in October 2013 Ebrahim (2013) found a flaw in twitter, called an unrestricted file upload. An application from the twitter developer centre (https://dev.twitter.com) is embedded with permission for the authors of the application to upload an image and link it to the application. This act bypasses authentication and uploads .htaccess and .php files to twing.com server, which can be used by criminals as Botnet that spread malicious contents. However, Zi Ciu et al (2010) found that between the human and the normal bots, exist what is called cyborg. It sends 'tweet' on behalf of the user and is activated right from the registration state where user set automatic programs to post twists during his/her absence. Twitter is vigilant in identifying flaws and fixing them, users especially in Africa should be aware of such security occurrences for a more proactive twitting habits.

Twitter has added more security features to make it harder for hackers to hack user accounts. This has become necessary for twitter with the hacking of high profile accounts, including that of Associated Press. With the added feature, called two-step authentication, in every login, a user enters a six digit code that is sent as a text message to the user's phone. This is to prevent password crack, with the introduction of the two levels authentication (something a user knows and something a user have). The causes of the account compromise are related to malware – phishing attacks, virus, and a user in an open and unsecure environment as well as the use of weak password. Although two factor authentication cannot completely solve the problem, yet stopping the use of applications like Tor could remarkably have an impact. Tor is a network that allows anonymous internet usage. It hides the online activities and location of the user from surveillance and traffic analysis. Moreover, basic settings in the twitter page that are often overlooked by the user do have a significant impact on user's privacy. For example, anything a user types in the box "what is happening," is seen by the public even by people without twitter account. The public display of user events can be hidden by going to settings and select "hide my tweet." Similarly, the applications the user interact with on the phone have access to the user's information.

5.5 The Facebook Security

As the computing world moves to mobile, the delivery and usage of social media platforms, such as Facebook, has shifted as well. More than half of the Facebook users look up Facebook on the phone. Facebook, as one of the giant and popular social media is making its presence in mobile devices. Facebook is turning human race into a single nervous system which will embrace every human mind. Thus, it is turning African society into connections of people into a single network in which all thoughts, plans, dreams, and actions of the Africa person will flow as nervous impulse in the network. The African person grew up to consume social media, but in today's electronic world, Facebook is consuming the African person. The personal and social activities of the mobile user are shared among friends, relatives and of course the hackers. Thus, a mobile user is put under surveillance, is monitored and tracked by the social media and the hackers. There are a lot of data coming out of Facebook, thereby giving away privacy of the user to third parties. In the Facebook page exist various applications in games and entertainments. The Facebook mobile operating system is not controlled by Facebook, but by Apple and Google. Most attacks now shift from OS to application programs. Third party applications are hosted on Facebook and this can endanger the user's privacy. Most of the applications are malicious and can steal personal details of a user and send the compromised account to the creator of the malicious application. This is known as identity theft and is common in Facebook. The same malicious application can disguise as a game; and in the process of playing the game with friend, malware is spread to the contacts of the infected friend. Facebook terms and condition contain a warning that this could actually happen, but users due to some reasons overlook or forget to consider the warning. There are malicious programs that allow the cracking of a Facebook account. Users should ensure that privacy settings are properly customized by hiding addresses and emails because they often serve as primary tools for cracking the account. Anything a user post online is seen by the public. The user should think before posting. Bad people can use the information or pictures of the user and post in other websites for nefarious motives.

Users should be more cautious on their Facebook activities for it is a platform that brings people of different characters some of whom are opportunity seekers to compromise the user's account. In taking the necessary precautions, users should begin setting the Facebook privacy from their Timeline. It is the Facebook latest update about user's profile. It puts all the stories, posts, and apps in one place. Users' should watch the video introducing Timeline on www.facebook.com/timeline. Users can choose to preview the timeline from time to time before friends can see what they post. This will give user time to decide what stories or photos to keep or delete on the timeline, and as well control who should see what. Similarly, in the section of privacy settings, user can control how much information to share with friends. Details of it can be explored for the user to get familiar with the various options available in the section. Below the privacy control is the default privacy control that is made up of three options: public, friends, and custom. The public means everyone on Facebook even if the person is not a friend can still see the post. The friend option means only friends can see the post; and the custom means the post is allowed for few selected people to see it or become hidden to another selected group of people. There are various simple to follow instructions while attempting to optimize privacy on the Facebook control settings. User should explore each settings to get familiar with them. Another control that is vital and crucial is that of apps. Applications can be a huge security risk on Facebook because they often need access to user's personal information. Users should take time to look at each application to know the type of information it is accessing.

5.6 Security in Mobile Social Media

The use of social media on mobile devices in Africa is increasing exponentially like wild fire. It is becoming the dominant life of the African people. Social media is the biggest transformation of African society since the industrial revolution. Social media is all about people talking to each other on the platform of software or program popularly known as social media. This social media include the Facebook, Twitter, Flickr, Ning, You Tube, dd.icio.us, Wiki, Blogs, yammer, LinkedIn, Mashups, and even emails and SMS. People use these media to interact, connect, and learn. People share knowledge, experience, information, and develop personally and professionally. Thus social media has become an integral part of African citizen and has influence on every aspects of his/her life. With the time and attention of people shifting to social media, hackers no longer spend tedious process and longer time trying to crack a system. Instead, they lure social media players (or users) to assist them in executing their dubious acts through phishing and malware. Karspersky Labs (2009) found that social media sites spread malware more rampantly than the traditional email phishing. The data that is collected and shared across the web in social media is so voluminous that hackers no longer have to search for victims, passwords, and leading information. All these are readily available to hackers and with little analysis their malicious acts is facilitated. The speech, interactions, and social life of users of social media are endangered with various forms of viruses and malware. According to Felt et al (2011) (Felt, Finifter, Chin, Hanna, & Wagner, 2011), the propagation of malware in social media is increasing in sophistication and strategies. Among the consequences malware is causing, as identified by Chiang and Tsau (2011) are battery depletion, exposure of user privacy, financial loss and data leakage. The design of mobile devices was not primarily intended to address security issues. As Zhang (2011) asserted, the primary concern for mobile device was optimization of battery consumption and device portability at the expense of security. Thus user's privacy was naked without protection and even the passwords and log in information go across the web in plain text. Thus, paving the way to compromise the user's data and cause havoc to the device functionalities. This coincide with the findings of Sophos (2012) and MacAfee (2012) who established that users store s and access sensitive corporate information on their mobile devices.

5.6.1 Mobile Malware

Malware is blended threats each designed to cause specific havoc on the user and the device. Virus, Botnets, spyware and adware are various form of malware aimed at compromising the user's privacy and device. With the proliferations of malware, malicious applications may be downloaded from various application stores or through files sent by MMS, SMS, Bluetooth, Wi-Fi, email, infrared, desktop Sync, or virtually any other means of import. Applications with malware infections that request permission could lead to an authorized transfer of information and money through initiation of phone calls or SMS messages, without user's involvement or knowledge. Another way malware is activated is through phishing. Phishing is an unsolicited junk messages or spam. It is common in social media, particularly in twitter. A user and a post can either or both be a spam. A user spam is written with curiosity terminologies, like: "have you seen the picture of yours and what was written on the blog?" This message is then followed by a link. If a user click on the link, then the user's account will be compromised, and send messages from a user's account without his/her knowledge. A user can do some basics checks to verify a post. First, a user can check the profile of the sender to see whether it is complete or poorly done. In most cases, incomplete profiles are spammers. The background and the link color should also be checked to see if they have been changed. However, some users may not have the time to customize some settings. Yet, it is worthy of verification. Another thing to watch out is if there is enough followers or twists. The more, the less likely, and the less the more likely it is a spam. If there are so many tweets, a user can usually see it as a stream of similar twists. Spammers are not conversational, they keep to one or few tunes and broadcast it over and over again. A user should be cautious in clicking a link, an advert or an image. Various strategies that exist to lure twitters into traps of spammers can be counteracted as Vathumasi et al (2012) (Vasumathi, Vaibhav, & Minaxi, 2012) asset that regular habit of verifying twitting events on the part of the user is desirable in the identification and prevention of the evolving complex spamming strategies.

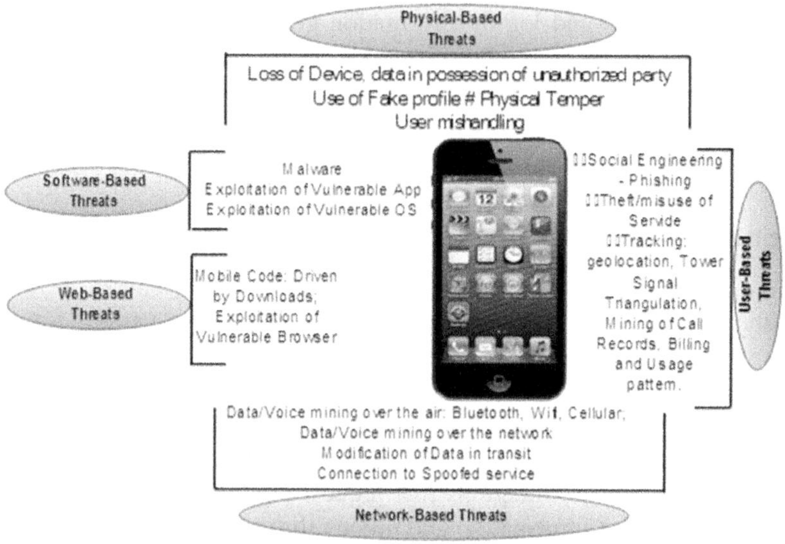

Figure 5.1 Mobile device vulnerability vectors

Mobile applications and mobile web browsers are the most commonly used applications to visit social media sites on mobile devices. Nowadays most of the social media sites have specific applications for mobile social media. Hackers too are not left behind in the provision of applications specific for mobile devices in social media. According to (Bhatti, 2012), the malicious applications of hackers are also present in the application stores of giants like Apple, iPhone, and Google Android. The user of bogus or fake profile, disguise applications, misleading links, Trojan software, illusive stories, and fake requests are very common with social media. The contents of the deception use thrilling messages, curiosity-based stories, and spoofed URLs to deceive users into clicking a link or downloading malicious contents. Similarly, the pool database of users in social media facilitate targeted attacks, where users receive messages from known friends; and based on trust, comply with the messages' requests thereby ending up as victims of malware and phishing attacks. The compromised mobile devices then become attack proxies a Botnets and DDoS (Distributed Denial of Service Attacks). Moreover many of the social media applications store the users contact address as the application is downloaded as well as the location and other identities of the user. The privacy of the user can be shared among different platforms thereby exposing the user to data harvest and location tracking. Other areas of mobile vulnerabilities include stolen or lost mobile devices, unencrypted storage of data, virus and spyware, remote access by hackers, and user interactions habits by the mobile device.

5.6.2 iOS Vulnerabilities

There is a security concern for iOS with the explosion of mobile applications in the African market. iOS security mainly depend on four pillars: device, data, network, and applications. The device pillar is supposed to prevent unauthorized access to the device through theft and lost. The data pillar is supposed to protect the data stored on the device, including unauthorized installations, modifications or usage of the data facilities. The network pillar is supposed to protect data in transit, including redirection of communication. There are flaws that make iOS, like iOS 7, vulnerabilities to hackers. For example if the phone is stolen, the find my phone features can easily be disabled by putting the iPhone or the iPad in the airplane mode. Similarly, the lock screen can be bypassed. This allows access not only to currently opened apps, but the camera role as well. The camera role internally grants access to any social media site. Simple pressing techniques on the phone provide access to the camera role where multitasking menu can provide further access to other apps. Similarly, the proliferations of payments in African market has promoted the user of applications that run on iOS for transport security for online banking and online trading. Most of the applications use SSL with https:// links. In terms of connectivity, iOS 3 and beyond has neat Wi-Fi hotspot feature that is set be default. As soon as a user connect, the first thing the iOS does is to go to www.apple.com and if it gets 200 response back, then it established a good connection; and if it gets two or three responses, then it decides to use the Wi-Fi spot and bring down the browser for the user to log in. Through this, if a hacker is in the same LAN with the user, the hacker can knock off the iOS device off the network when it re-joins. The attacker then forges it to redirect to a malicious website.

Thus, iOS attack is multi-faceted that can by anything from lost/stolen device, physical access and installation of malware via browser JavaScript engine, and apps downloads from enterprise app store. Another vulnerability with iOS 7 is the call bug attack, using the phone at freezing mode to make calls, updates Facebook and twitter status, or send messages. All these can be done without log in to the pass code. For example, if a user left the phone unattended, an intruder with the knowledge of the flaws and the techniques, could infiltrate the privacy of the user. Bypassing the pass code and the lock screen of the phone is a serious security concern. The lock screen is the most fundamental and most obvious security level on the phone. Security is tight up with the lock screen. If it is defeated, then all layers of security are rendered defence less. User can try sequence of interactions with the phone at different trials to discover whether his/her phone is among with such flaws. If it is, patches and updates with relevant applications on the manufacturer's website could fix the issue. The trial is essential because the revelation of the flaws may encourage criminals to apply the findings on other devices. If by chance, similar flaws or oversights are found, then they get control of the device and its functionalities.

5.7 Possible Solutions

Although solutions were mingled in the discussions of previous malicious events, yet specific strategies for mitigating issues of mobile security is desirable. Against this background, the discussions that follow offer possible solutions that could address the issues raised and provide security outline within which users can operate with a sense of safety. Mobile service providers in partnership with manufacturers, vendors, and service agents should create a system and framework that would constantly and regularly teach, train, educate and update users on the dangers they may be encountering on their mobile devices as they socialize and interact on the day to day basis on their phones. Such a framework should include a curriculum that step by step structured how a user can differentiate the normal and abnormal device behaviour in terms of operations with the device and interactions with applications. Ability to notice unusual behaviours will give the user a feeling of an attack or the presence of malicious contents on the device. How to recognize and identify malicious applications should also be among the areas the user should be exposed to. Through peer review of mobile applications, users can see the ratings, comments, and experiences of other users with various applications; and that would help in furnishing user with the information necessary to decide on a particular application. A user should do a little research before downloading an app to ensure that it is from well-known source so as to avoid downloading malware that may monitor keystrokes or reporting on user's activities. If a user has already installed a program and found it to be malicious, the process of uninstalling and deactivating the program should also be stated through settings and removal of programs. The completion of customization of privacy settings should be properly done before acknowledging the signing up with a social media. This takes care of any vulnerability that might be opened due to misconfigured settings. The mobile browser settings should be customized and optimized to close loop holes for vulnerabilities and exploits.

Mobile browser known as micro browser, mini browser or wireless internet browser is the web browser designed for user on mobile devices such as a mobile phone. Mobile browsers are optimized so as to fit web contents more effectively for small screen and portable devices. Most smartphones and tablets computers come with a built in mobile browser for the web. The most popular browsers are Google chrome, IE, Mozilla Firefox, and Opera; which is fast and uses as little as one tenth of data, thereby saving on data cost. Each browser has its own peculiar vulnerabilities and peculiar settings options. A user should explore the browsers frequently used and get familiar with the settings for privacy and safer browsing. Similarly, the applications on the devices need to be updated regularly and frequently. This is necessary because all software has bugs, which are vulnerabilities and security holes. The software and applications on the user's devices may have unknown vulnerabilities. Immediately criminals are aware of it, they rush to develop applications that can compromise user's privacy and device. Users should be vigilant and active in updating the applications running on their devices. User security consciousness and habits in terms of passwords, device handling, and information storage, should be proactive and smarter. User should create strong password that is hard to guess or crack. It should be strong enough that even in the event of losing the device, the device can be useless to the thief or unauthorized intruder, physically or remotely. Similarly, the user should ensure that the device is locked up every few seconds and tolerate the efforts of re-login with the difficult and hard to guess password. User should keep as little information as possible and ensure that the information is backed up.

Moreover, mobile platforms in most cases, are not integrated with security features. This give room for hackers to exploit the devices and take control of the data and the devices. Against this background, telecom providers should endeavor to ensure that necessary provisions are made to protect the devices, user privacy and data. The major beneficiary of the growth and sustainability of mobile devices are the telecom providers. The higher the turnover in smartphones and data friendly phones, the higher the profitability and growth of the telecom company is. If users avoid the use of smartphones due to exploit and vulnerabilities, the more it affect the telecom companies financially and competitively. Users consume more data than on calls services. It therefore becomes desirable for telecom companies to play a major role in the security of mobile phone in user privacy, and malware attacks on mobile devices. User's data must be protected whether in motion or residing in their SIM card or

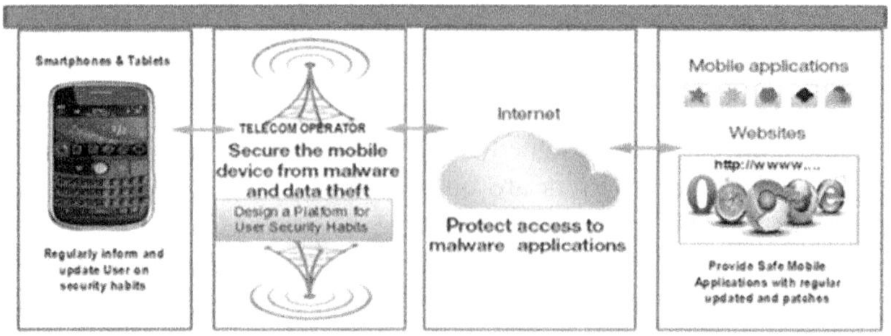

Figure 5.2 Telecom provider role in mobile security

phone memory. Unauthorized access to user's data should be one of the focus areas of security by the telecom operators. A framework that would track, trace and monitor stolen phones should be deployed to give user the confidence and security to buy more expensive phones that could enable them to use more data for social networking and related data usage activities implying more business and profit for the telecom company. All these should be embedded and integrated in the customer service. There should be a paradigm shift in the perception of telecom company as per customer service is concerned. The traditional customer service is to provide assistance and support to service functionalities and implementation, and problems. With the evolving threats in mobile devices, service delivery and quality of service are becoming more crucial and deserve special attention. In this regard, Telecom Company should play an active role in the actualization of security objectives of confidentiality, integrity and availability for the user. If for instance, a user's is stolen the telecom operator should remotely locate and lock the device. The telecom operators should have security section purposely for security customer service that is proactive in emerging threats and proper user security behavior. The diagram below shows how such role can be structured.

The telecom operators must be involved in the mobile security controls, they should not be common carriers of data or pipe for tunneling information. The user is the catalyst for the growth, survival and profitability of telecom services. A satisfied user is not only a loyal customer but a network that can sell and advertise the services of a telecom company. Likewise a dissatisfied customer is a network that spread negative messages, experience and encounters with a telecom services. Quality of service is one of the significant area in customer service. With the increasing popularity and use of social media like the Facebook and Twitter, coupled with multimedia applications, users spend more on data than on voice calls. This is further intensified with the penetration of smartphones in the African society. For the user to have a return on investing on data to enjoy social media services, the user should have satisfaction in terms of privacy, integrity and availability. If at any point a user encounters privacy intrusion, identity theft or connectivity bottlenecks due to Denial of Service Attacks (DoS), the user may feel frustrations with data usage applications. The user may suspend or stop spending on data usage, and that in turn affect the turnover of the telecom company. Data consumption is really the driver of business growth of telecom provider, and any challenge to that must be faced with all possible strategies at the disposal of the telecom operators. In order to stop this from happening, telecom companies should be proactive in providing security measures that provide safer usage on data. In the previous figure the telecom operator should regularly inform and update users through SMS to become active in the updates of the applications mostly used in the user's device. The company should ensure that the process of the update is facilitated for the user through easy to follow instructions. Short SMS on proper security habits should be sent constantly to users, with highlights on the dangers and threats confronting an abuse in proper security habits. Nowadays malware vulnerabilities are targeting applications rather than operating systems of the devices because of the pooper security habits of the end user. The ISP should therefore focus more on user awareness campaign. This process would provide a protected and secured mobile device user. The telecom operator should bear the responsibility of detecting and notify users of possible malware infection, and educate users on how to be vigilant and sensible.

5.8 Conclusion

Threats to mobile devices in terms of vulnerabilities and attacks, if not checked and controlled, could have serious setback to the growth and development of communication in Africa. The African market is very large and growing every day in terms of consumption and purchase of mobile phone, data and connectivity services. The fast growing technology in Africa is mobile phones. There are more than 800 million phone users in African countries south of the Sahara desert. This outnumbered the mobile users in Europe and the US. The data location service exposes the physical whereabouts of a user. Such services should disabled so as to avoid being exposed or tracked anytime an update is posted on the social media. Mobility of mobile phones and services is being confronted with security challenges that is not only affecting the user but significantly affect the telecom operators and service providers. User has the option of resorting to low level entry phones if the issues of security on mobile devices continue to become complex. If this could happen, it will be a great set back to the telecom operators. The consumption of data services by mobile device users will drop, and this implies a drop in turn over. This in turn affects; growth, survival, and profitability of the telecom operators. The consequences could spread to other sectors of the economy and social life of the people. Smart phones are increasingly becoming more complex and the complexity gives rise to all sorts of vulnerabilities. Apart from user awareness, telecom companies in collaboration with mobile apps developers should start looking for ways to minimize the vulnerabilities, threats and attacks so as to provide proactive security on mobile devices.

Among the vulnerabilities African mobile user is encountering are: stolen or loss of devices; privacy intrusion, social engineering, and hacked device. Vulnerability also occur at connectivity through Bluetooth, in which an unauthorized connection is made to steal user's data and intrude on privacy; through Wi-Fi, in which user is redirected to malicious website; through 3G or 4G in which user is lured into clicking links and attachments downloads from malicious websites; through GPS in which user's whereabouts is tracked and monitored. Similarly, unauthorized access to user's devices through malware in apps and website interactions are common with mobile devices. Phishing through emails, text message, ads, and apps are other vectors of vulnerabilities in African mobile devices. Users should be sensible in responding to messages with offers and benefits. Users should run reliable and trusted apps in devices. Antivirus software should be installed and constantly updated.

6

Achieving Scale and Sustainability in M-health Solutions for HIV/AIDS in Africa

Perpetual Crentsil
Social and Cultural Anthropology,
University of Helsinki
crentsil@mappi.helsinki.fi

6.1 Introduction

The advances in mobile technology have generated an upsurge in the use of mobile phones to improve health outcomes, which is known as mobile health, or m-Health, and is a sub-segment of the broader field of electronic health (e-Health). Mobile phones have shown promise in providing greater access to healthcare to populations in African countries and especially in HIV treatment and prevention initiatives. The use of mobile phones to improve the quality of care and enhance efficiency of service delivery within healthcare systems has created cost efficiencies and improved the capacity of health systems to provide quality healthcare (Lemaire, 2011, p. 10). The World Health Organization (WHO) has defined m-Health as the "provision of health services and information via mobile technologies such as mobile phones and Personal Digital Assistants (PDAs)" (WHO, 2012).

Donner (2008) outlines three major trends in the studies of mobile phones in developing countries—the adoption of phones, impact of the devices, and the interrelationships between mobile technologies and users (p. 141). The mobile platform presents a unique capability of delivering healthcare services wherever people are—not just in healthcare facilities. M-Health initiatives have also been effective in reaching underserved populations, particularly those in rural areas, to change health behaviors and outcomes. The devices address a wide variety of healthcare challenges, including the shortage of skilled healthcare workers in rural areas, treatment adherence and compliance, lack of timely and actionable disease surveillance, lack of medical diagnostic treatment, and slow rates of information flow and reporting delays (Lemaire, 2011).

Yet, scaling up and sustaining these programs have less been addressed. Despite hundreds of m-Health pilot studies, there has been insufficient programmatic evidence to inform implementation, scale up, and support sustainability of m-Health solutions in HIV/AIDS initiatives in Africa (Tomlinson, Rotheram-Borus, Swartz, &Tsai, 2013). While m-Health projects in the HIV/AIDS sector provide a solution to the limited reach of existing health service infrastructure in most sub-Saharan African countries, the programs face sustainability and scalability challenges of their own. The problem remains how projects can be scaled up, the institutional arrangements, and how various stakeholders can co-operate to provide applications in affordable and sustainable ways (Tenhunen, 2013), in order to help AIDS patients and those most affected by HIV/AIDS in Africa.

There are over 600 million mobile phone subscribers in Africa, out of the world's 6 billion. In 2012, Ghana's rate of mobile phone penetration was 100 per 100 inhabitants—26.3 million subscribers as in February 2013 in a population of 25.3 million.[1] South Africa's rate of mobile penetration was 135 mobile subscribers per 100 inhabitants in a population of over 50 million and 68.3 subscriptions. Uganda had a mobile penetration of 46 per 100 inhabitants in its population of 36.3 million with 16.3 million subscriptions. It must be understood, however, that users might have multiple subscriptions while others might have none. The mobile phone penetration rates have created huge excitement and the possibility of reaching and following patients who were previously unreachable via traditional communication channels. The WHO recommends continuing mentorship via face-to-face training, telephone calls and other information communication technology (ICT) approaches to support scale-up of HIV care, anti-retroviral therapy and prevention in resource-constrained settings (WHO, 2005, p. 34). Consensus is that m-Health represents a cost-effective technology solution to many of these challenges if implemented correctly, brought to scale and sustained (Lemaire, 2011), especially in the HIV/AIDS sector in Africa.

Sub-Saharan Africa remains the most heavily affected region in the global HIV epidemic. According to a United Nations Program on AIDS (UNAIDS) report in 2011, an estimated 23.5 million people living with HIV were

---

[1] http://www.itu.int/net/pressoffice/press_releases/2013/CM12.aspx

resident in sub-Saharan Africa, representing 69% of the global HIV burden (UNAIDS, 2012). And new infections continue to occur. The major mode of HIV transmission in much of sub-Saharan Africa is heterosexual contact and mother to child infection (ibid). While knowledge and awareness of HIV/AIDS and its proven modes of transmission has increased over the past few years, this knowledge is often intertwined with myths outside of the biomedical paradigm; beliefs are rife that an infected person having sex with a virgin can be cured of the disease as well as supernatural ideas like witchcraft, sorcery and curses as the cause of infections (Radstake, 1997; Crentsil, 2007). These beliefs can influence risky sexual lifestyles, when people feel protected spiritually and therefore do not see the necessity to undertake preventive measures.

International and national policies have been in place for prevention alongside treatment. Many national organizations in sub-Saharan Africa, with support from international agencies, are striving to provide antiretroviral treatment (ART) to patients in need (Bärnighausen et al., 2011, p. 942). A joint project developed by the UNAIDS, the United States Agency for International Development (USAID), WHO, and the POLICY project recommended the AIDS Program Effort Index (API) as a tool for monitoring the 2001 Declaration of Commitment to HIV/AIDS by measuring high-level program inputs in the worldwide fight against the disease (Bor, 2007, p 1587). Agencies involved in HIV prevention strategies in Africa have made considerable use of mass media (Benefo, 2004), including radio, television/cinema, newspapers, billboards, and posters as means of exposure and education and have also been one of the most important predictors of awareness and behavior change. Mobile phones and other ICTs have become important communication tools used in African healthcare for knowledge dissemination, as alert systems and in interactive formats, particularly in HIV/AIDS initiatives (Chib, Wilkin, & Hoefman, 2013).

At its most basic level, the concept of using ICTs to propel health and development is a paradigm which presupposes that the availability of, and access to, ICTs will incontrovertibly lead to development (Han, 2012; Phippard, 2012). The integration of m-Health in HIV/AIDS initiatives has demonstrated quite positive outcomes, but the sustainability and scalability operations require a closer attention and evaluation. The policy infrastructure for funding, coordination and guiding the sustainable adoption of m-Health services also remains under-developed (Tamrat & Kachnowski, 2011). Applications of mobile phone technology in HIV and other disease prevention efforts have been along the lines of health promotion in general (Mitchell, Bull, Kiwanuka, & Ybarra, 2011, p. 771).

Using my own ethnographic research in Ghana and an intensive literature review, this paper explores experiences in implementing, scaling up, and sustaining mobile phone-based health projects in HIV/AIDS initiatives in Ghana, South Africa, and Uganda. The three countries have been chosen because they have a number of m-Health projects and experiences that offer good examples of both potentials and challenges of mobile technology in the HIV/AIDS sector in Africa. This article aims to explore how a lasting impact at scale and sustainability can be made in m-Health for HIV/AIDS in Africa.

6.2 Scaling Up and Sustainability of Health Promotion Programs

Scaling up and sustaining health promotion interventions are very important in projects and programs. Swerissen & Crisp (2004) maintain that knowing what it is that one seeks to sustain is a useful start, but whether such aspirations are realistic is another question (p. 123). Sustainability entails development, and involves strategies for assessing what benefits need to be sustained over what time frames with what resources. In the health promotion literature, there is considerable concern about the need to maintain and retain health promotion projects and programs long term (Swerissen & Crisp, 2004). In an editorial comment, Leger (2005) claims that although many project plans usually contain the word 'sustainability', many of them do not tease out those important aspects of the intervention which are worth sustaining. "They do not even identify whether the intervention itself actually nurtures the process necessary to ensure the stated intentions have a reasonable chance of being sustained" (p. 317).

Swerissen & Crisp (2004) have pointed out that the aim of health promotion is "to produce intervention effects that may be sustained over time" (p. 123). Recently, a number of researchers have sought to interrogate the concept of sustainability in health promotion, and to develop frameworks and guidelines which will assist planners, practitioners, evaluators, managers, organizations, and funding agencies (Leger, 2005). It has been suggested that health promotion and development theory must involve multilevel analysis of patterns of change, and away from only one mode of problem solving. Swerissen & Crisp (2004) define sustainability in health promotion as the intervention effects or the means by which they are produced by the programs and agencies that implement interventions (p. 123).

A number of studies have identified numerous factors related to program longevity, including, among others, presence of a program champion, involvement of key community leaders in a programme development and implementation, staff involvement in decision-making concerning the programme, and fit of the programme to the organisation's values and norms (O'Loughlin, Renaud, Richard, Gomez, & Paradis, 1998). Social intervention theorists have proposed a look at the social order of a society, and Swerissen & Crisp (2004) have examined sustainability of health programmes within different levels of social organisation. The authors suggest three key factors for the development of a policy on sustainability for health promotion. These are: levels of social organisation which are the focus for change, programmes and agencies which are the means employed to achieve change and the outcomes or effects that are achieved (p. 124). They propose a typology where there are four levels of health promotion intervention; that is, at the individual level, organisational, community action and institutional. Programmes, agencies, and/ or effects need to be explored fully for issues surrounding the differing targets for scaling up and sustainability (Crisp & Swerissen, 2002).

Earlier, Shediac-Rizkallah & Bone (1998) had examined indicators of sustainability and they saw the need for thoroughness in planning for sustainability, which requires, among others, a clear understanding of the concept of sustainability and operational indicators like level of institutionalization and capacity building in the recipient community. They proposed three categories for exploration— individual health benefits, institutionalization factors, and community capacity attributes. Leger (2005) has even been bold to suggest that many health promotion initiatives that were largely individualistic and behaviorist have failed to be sustained.

In the late 1990s, O'Loughlin et al. (1998) examined the correlates of the sustainability of community-based heart health promotion interventions in Canada. They found that achievement of community-wide improvements in lifestyle habits through the implementation of health promotion programs can be a lengthy process because large segments of the population must be exposed to the program; among those exposed, persons in need of change must be exposed at a level of intensity and duration sufficient to produce behavior change. Again, they found that interventions that underwent modification during the implementation were almost three times more likely to be sustained than those that remained in their original format. Also, the quality of the intervention-provider fit was associated with sustainability, and finally, the presence of a program champion who strongly advocated the continuation of the intervention was important to sustainability (p. 707).

Recently, scaling up and sustaining m-Health projects and programs have gained interest in research, and increasing efforts are being directed toward addressing the sustainability of such m-Health interventions (e.g., Sanner, Roland, & Braa, 2012; Leon, Schneider, & Daviaud, 2012; Tomlinson et al., 2013). M-Health program sustainability has also become an issue of growing anxiety as policy makers and funders become increasingly concerned with allocating scarce resources effectively. Leon et al. (2012) have applied a framework for assessing the health system challenges to scaling up m-Health in South Africa, based on three factors: organizational systems, technological systems, and financial systems. They concluded that where a health system has a weak ICT environment and limited implementation capacity, it remains uncertain that the potential benefits of m-Health for Community-Based Health Services (CBS) would be retained with immediate large-scale implementation. Many barriers to access and usage hinder many realizations of m-Health projects. In a study of m-Health for HIV/AIDS solutions in South Africa, Phippard (2012) identified potential barriers including gender, language, literacy, training, financial sustainability, scalability, government cooperation, and dependence on inadequate pre-existing health infrastructure.

We need to understand new media and social change through exploring how different forms of mediations interact as part of local hierarchies when a powerful new medium is appropriated, but we also need to examine scale up and sustainability of mobile phone use for HIV/AIDS measures in Africa. In doing so, I also look at the specific communicative ecology related to HIV in Africa (Ghana, South Africa, and Uganda). Communicative ecology, defined as the dynamics between human communication and the effective environment, is primarily understood as the link between human communicative processes, structures and meanings, social networks, and communication technology (Foth & Hearn, 2007). People interact with others through communicative structures that are mediated, informational, situational, and contextual in their respective environments (Matthias, 2011, p. 31).

6.3 Methodology

The analyses in this chapter are based on Swerissen & Crisp's (2004) classification of scale and sustainability within different levels of social organization as well as the framework used by Leon et al. (2012) and adapted from three approaches to reviewing sustainable ICT solutions (organizational systems, technological, and financial systems). This paper applied most of these factors and others that affect the scale up and sustainability of m-Health solutions in HIV/AIDS initiatives in Africa.

6.3.1 Materials

This chapter is based on an ethnographic study of mobile phones and healthcare communication in the HIV/AIDS sector in Ghana. It also relies on an intensive review of literature on implementing, scaling up, and sustaining mobile phone-based health projects in HIV/AIDS initiatives in Ghana, South Africa, and Uganda. The ethnographic data were collected over five months from 2010–2011 through qualitative methods of observation and semi-structured, face-to-face interviews with health personnel, HIV/AIDS counsellors, HIV-positive people, and mobile phone users in four towns and villages (two urban and two rural) in southern Ghana. The four research sites had one urban and one rural area each with a hospital and an HIV unit where I could interview people and observe HIV/AIDS patients and health personnel for mobile phone health care communication. The sites were selected so that the study would incorporate data from different social environments in southern Ghana and thus ensured that the data was as representative as possible of the situation. In addition to these materials, I draw on my prior research in Ghana. As a Ghanaian and having conducted several periods of fieldwork in the country since 1999 (see Crentsil, 2001, 2007), I am familiar with healthcare forms, the social impacts of HIV/AIDS, and the place of mobile phone technologies in the communicative ecology.

This chapter is also based on a review of 28 publications that related to scale up and sustainability of m-Health solutions in Africa and retrieved electronically between 1 August and 31 October 2013 in 2 databases: Google Scholar and PubMed.[2] Articles focusing on the world, Sub-Saharan Africa, or Global South were included if they focused also on some African countries. Sources included published academic articles or reports, personal or institutional blogs on the internet and grey literature. Title, abstracts and descriptor terms of the electronic search results were screened independently for relevance based on the types of participants, interventions, and study design. The review is primarily peer-reviewed or edited publications: books/theses, journal articles,

[2] My search in PubMed was brief because I stopped it when I noticed that most of articles there were found in Google Scholar too. Eventually, I added only one article from PubMed to those selected from Google Scholar.

book chapters, and papers presented at conferences or in proceedings which utilize full paper peer reviews. The review is a sample, rather than a census, of documents and was limited to English language publications on the subject of scale and sustainability in Ghana, South Africa, and Uganda, published between 1 January 2010 and 31 October 2013, which yielded a total of 4,710 results (see Figure 6.1 below).

To identify discussions of scale up and/ or sustainability of m-Health for HIV/AIDS initiatives in Ghana, South Africa, and Uganda, terms that would incorporate m-Health and HIV/AIDS prevention and/or treatment topics, including keyword combinations of HIV, AIDS (or HIV/AIDS), "mobile", "technology", health, m-Health in Africa, ICT, telemedicine, "HIV prevention", "AIDS treatment" were searched. A further search was conducted using the terms "m-Health in Africa", and also "scale up", sustainability", "barriers", "challenges", "factors for using m-Health".

Not every text was reviewed; that is, I adopted an inclusive or exclusive criterion in order to identify only publications relevant to the aim of this paper. I reviewed, identified and retrieved all eligible documents. The titles and abstracts were first examined to remove obviously irrelevant reports. Articles were included when they related to HIV/AIDS interventions using SMS and/or voice calls in African projects, with a particular focus on Ghana, South Africa,

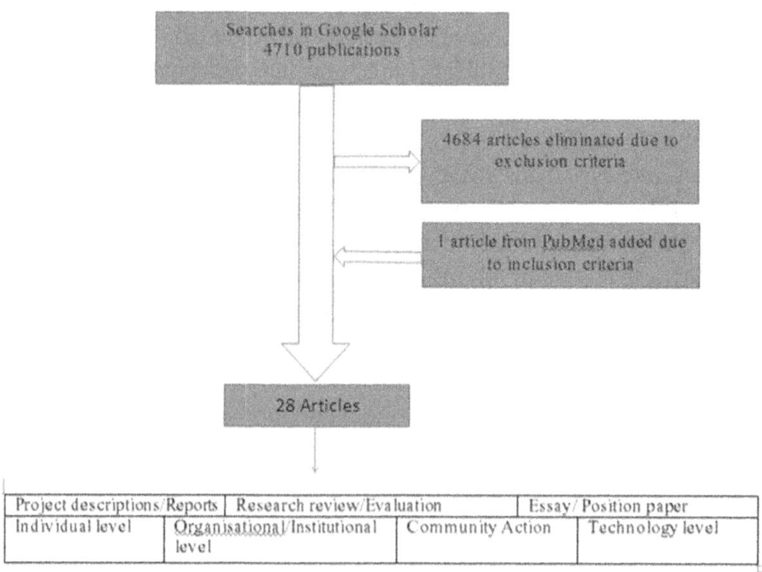

Figure 6.1 PRISMA Flow diagram of internet search for literature reviewed

and Uganda. However, other publications on other areas of health in other African countries were included. One article was based on tuberculosis (TB) in South Africa (Chaiyachati et al., 2013), HIV care in Kenya (Thirumurthy & Lester, 2012), and two others in Botswana on patient adherence (Littman-Quinn et al., 2011a) and on public health interventions (Littman-Quinn et al., 2011b) were included because they revealed scale up and sustainability challenges in m-Health solutions. Eventually, 28 articles were selected for review. For the purpose of this article, m-Health referred to all applications that used SMS or texting and voice calls for prevention and/or therapy purposes in HIV/AIDS initiatives in Africa.

6.4 HIV/AIDS in Africa

Sub-Saharan Africa is the region worst affected by the HIV/AIDS pandemic than any other region worldwide. The region bears the brunt of the epidemic and is home to about 69 per cent of the 33 million people infected with the virus globally, meaning that more than two-thirds of people living with HIV in the world reside in sub-Saharan Africa (UNAIDS 2012).

In 2008, the region accounted for 68 per cent of the 2.7 million new infections among adults worldwide (UNAIDS/WHO 2009), and the region accounted for nearly three-quarters of AIDS-related deaths that year (UNAIDS, 2009, p. 21).

The rate of HIV in women is generally higher than in men of the same age group (UNAIDS/WHO, 2009). Ghana's current prevalence rate is about 1.4 per cent in a population of 25.3 million. In 2012, around 240,000 people were living with HIV in Ghana, 120,000 of whom were women aged 15 and above.[3] Ghana recorded its first HIV/AIDS case in 1986 among sex workers who had returned from a sojourn in neighboring Côte d'Ivoire. A medium-term plan developed with the WHO's Global Program on AIDS in late 1988 set up the National AIDS/STI Control Program (NACP) in the Ministry of Health aimed at prevention, management, and control of HIV infection in campaigns through the mass media (Crentsil, 2007). South Africa's 2012 estimates were 6.1 million people (about 17.9 per cent in a population of about 51 million) living with HIV, of whom 3.4 million were women.[4] AIDS was first diagnosed in South Africa in 1983. According to a report for the National Commitments and Policies Institution (NCPI), South Africa developed a multi-sectorial strategy

[3]See UNAIDS, http://www.unaids.org/en/regionscountries/countries/ghana/

[4]See UNAIDS, http://www.unaids.org/en/regionscountries/countries/southafrica/

to respond to HIV/AIDS. The National Strategic Plan for 2012-2016 continues to tackle areas such as prevention, treatment, care and support, research, monitoring and surveillance, as well as human rights and access to justice (UNAIDS 2012). Uganda's estimated 1.5 million people living with HIV (a prevalence rate of 7.2 per cent in a population of 36.3 million) has 780,000 women.[5] AIDS was first identified in Uganda in 1982 in a fishing village near Lake Victoria (Tumushabe 2006). By 1986, the government had admitted that there was an epidemic and the various players—donors, government, non-governmental organizations, faith- and community-based organizations, and families all got involved in the struggle against the epidemic (ibid; Pool, Kamali, & Whitworth, 2006).

Despite the slight decline in overall HIV prevalence, sub-Saharan Africa accounted for the vast majority of new HIV infections worldwide among both adults (68%) and children (91%) in 2008; more than 14 million children have lost one or both parents to the epidemic (UNAIDS, 2009, p. 21). Women are particularly vulnerable to infection due to low condom use in Africa (UNAIDS, 2012) and the fact that casual sex, pre- and extra-marital relationships are common. In many parts of Africa, men are assumed to be inherently promiscuous and resisting the use of condoms. Men may marry more than one woman, and many married men indulge in one-time intimacies or extra-marital affairs with girlfriends (Moore & Williamson, 2003; Crentsil, 2007).

Educational campaigns about HIV/AIDS emphasize prevention; those already infected receive counselling and therapy with Antiretroviral drugs (ARVs). Most HIV prevention programs have been built on a paradigm of individual behavior change; people are advised to lead healthy sexual lifestyles by practicing "safe" sex grounded in the widely-known ABC method—*A*bstain from sex, '*B*e mutually faithful' to partners, or '*C*ondomize' (i.e., use condoms consistently). Despite widespread awareness and education, HIV/AIDS remains a stigmatized illness in Africa. AIDS patients are usually seen as having indulged in sexual (mis) behavior and, in religious contexts, are being punished for their "sins" (Crentsil, 2007). Secrecy, denial, and shame usually characterize the disease (Radstake, 1997; Crentsil, 2007; Kalichman & Simbayi, 2004).

One problem acknowledged by researchers and stakeholders in the HIV/AIDS sector is that for many patients, there is an unmet need for information on structures of support (De Tolly & Alexander, 2009,

[5]See UNAIDS, http://www.unaids.org/en/regionscountries/countries/uganda/

p. 1). In an article on mass media and HIV prevention in Ghana, Benefo (2004, p. 2) points out that government and non-governmental organizations (NGOs) alike have utilized mass media as a component of information, education and communication, (IE&C) as well as in social networking campaigns to disseminate information about HIV/AIDS, reduce misinformation and induce behavioral changes against risks of HIV infection. There is widespread consensus that ICTs present the best solution to this problem, with mobile phones showing particular promise in HIV/AIDS initiatives (Phippard, 2012).

6.5 Mobile Phones and M-health in HIV/AIDS Programs in Africa

The growing enthusiasm for m-Health is driven by the demonstrated benefits of mobile technology as well as by widespread availability of mobile phones and the relatively low levels of literacy required to use them (Crentsil, 2013a, b; WHO, 2012; Leon, Schneider, & Daviaud, 2012). Mobile health has been advocated as an innovative tool for improving both access to and quality of health care in low and middle-income countries (Chang, et al., 2011). Many health systems and projects in Africa have implemented mobile phones for health outcomes, to facilitate emergency medical responses, point-of-care support, health promotion and data collection in the HIV/AIDS sector. Mobile phones enable health practitioners and experts to reach and treat patients over a far wider geographical range than was possible in the past (van Beek, 2009, p. 133).

In HIV/AIDS counselling and therapy, the most common form of adherence monitoring is structured around patient interview, usually at health facilities and commonly referred to as self-report (Haberer, Kiwanuka, Nansera, Wilson, & Bangsberg, 2010). While this mode of communication has been effective, there are concerns of privacy and shame but m-Health seems to be encouraging patients to talk more freely about their statuses than in face-to-face encounters (Crentsil, 2013a, b, in press). Mobile phones can be used in very private ways with high anonymity, as well as with personalized access to sensitive information. Their flexibility, particularly the ability of making calls unobtrusively, is undoubtedly appealing for communication about the disease due to the stigma and shame surrounding HIV diagnoses. M-Health shows great promise as a means of enriching health communication across the entire AIDS continuum by "strengthening both the scale-up of prevention, outreach, and awareness programs and the access to treatment, care, and support for people living with HIV" (Phippard, 2012, p.178).

The general healthcare needs for information technology are increasing productivity, capacity and patient service, as well as the clinician's needs for providing the right care to the right patient at any time, in any place (Yu, Wu, Yu, & Xiao, 2006, p. 181). In Africa, mobile phones are important in social relationships and kinship networks by helping to maintain links with family, friends, neighbors, business customers, and health personnel (Crentsil, 2013a, in press), as Figure 6.2 below shows. Phone sharing with relatives and friends is common, and a mostly pre-paid card that allows a user to make calls until the budget has been depleted, after which the phone can only be used for receiving incoming calls until the card's expiry date (cf. Kaplan 2006), offers a low-cost use of the mobile phone.

Mobile health as the provision of health services and information via mobile and wireless technologies has become ubiquitous and made m-health applications an important tool with which to impact the health of people. When applied correctly, m-health can make real contributions to improved health outcomes. M-Health has the potential to address and overcome: (1) disparities in access to health services; (2) inadequacies of the health infrastructure within countries; (3) shortage of human resources for health; (4) high cost of accessing health; and (5) limitations in the availability of financial resources (McQueen et al., 2012). M-Health refers to the use of mobile devices, such as phones, to

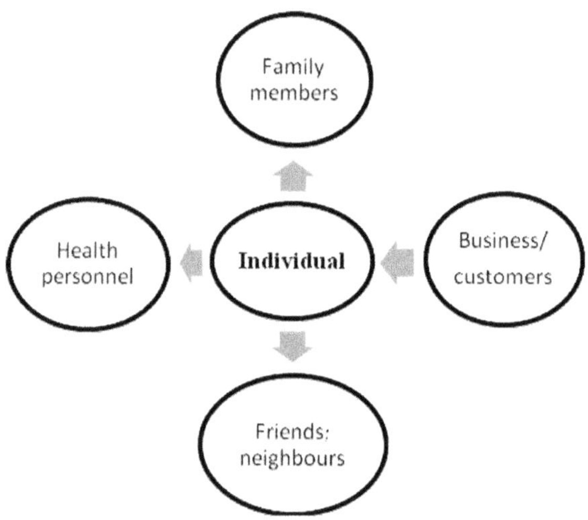

Figure 6.2 Communication with significant others

support the practice of medicine and public health; it is a rapidly growing field with potential applications for frequent self-reported adherence data collection (Haberer et al., 2010)

Many African countries have quite high mobile penetration. In Ghana, the rate of mobile penetration is 100 per 100 inhabitants in 2012. According to the World Bank, there are more than 27 million subscribers in Ghana's population of 25.3 million with a gross domestic product (GDP) of $40.7 billion and ranked 135 out of 182 countries in the United Nations Development Program (UNDP) Human Development Index. South Africa's rate of mobile penetration is 135 subscribers per 100 inhabitants and 68.3 million subscribers. With over 51 million inhabitants, South Africa is considered one of the most emerging countries, with a GDP per capita of $384.3 billion and ranked 121 in the Human Development Index. Uganda has a mobile penetration of 46 per 100 inhabitants and 16.3 million subscribers in its 36.3 million population, a GDP of $19.88 billion and ranked 161 (World Bank , 2013).[6]

Mobile phones are being used quite extensively by HIV counsellors to deliver key information to patients faster and without transport costs, and are seen to help avoid the stigma and shame often associated with being HIV positive in Ghana because they can ensure secrecy and anonymity. Mobile phones are slowly being used to deliver information to other areas of healthcare services. There is currently one mechanism in place to solve the problem of fake medicines by sending a free SMS text to a central number to check the authenticity of a drug. Another example is the One Touch Medicare line, which provides doctors with free mobile phone calls and text messages to other members of the medical community nationwide (Luk, Zaharia, Ho, Levine, & Aoki, 2009). The USAID/Ghana Strengthening HIV/AIDS Response Partnership with Evidence-Based Results (SHARPER) project, led by the Family Health International (FHI) 360, aims to contribute to Ghana's national goal of reduction in new HIV infections by fifty per cent by 2013 through delivery of an integrated project, tailored to the needs of key populations and their partners. Launched in September 2008 under a previous USAID-funded project, the Text Me! Flash Me! Helpline uses cell phone technology to provide Most at Risk Persons (MARPs) in Ghana with friendly and accessible HIV/ AIDS and other sexually transmitted infections (STIs) information, referrals, and ART reminders through SMS and counselling services from qualified providers. The target audiences are MARPS—men

[6]Also UNDP–http://hdrstats.undp.org/en/countries/profiles/UGA.html

who have sex with men, people living with HIV/AIDS, and female sex workers (McQueen , et al., 2012)

South Africa has even a more promising scenario and the use of SMS texts to expand the uptake of HIV testing and follow-ups in treatment are wide-spread; for example, the Cell-Life project and others use mobile phones as a mass information channel for HIV/AIDS interventions (www.celllife.org; De Tolly & Alexander, 2009). Cell-Life is also developing solutions that support the management and monitoring of HIV/AIDS in partnership with the University of Cape Town, the U.S. President's Emergency Plan for AIDS Relief, USAID, Vodacom, and other international organizations. Solutions include the intelligent Dispensing of Antiretroviral Treatment for adherence improvement, the data collection tool Aftercare, and the use of SMS texts to expand the uptake of HIV testing and follow-ups in treatment. Similarly, Project Masiluleke, also in South Africa, is structured around the use of mass HIV/AIDS awareness-raising and educational messages that are inserted into the unused space at the bottom of a specialized text message, "Please Call Me." MTN, South Africa's mobile phone provider, donates 5 per cent of the space for use by Project Masiluleke, and one million Please Call Me messages are sent every day to South African mobile subscribers (Phippard, 2012; PopTech, 2012).

In Uganda, a number of pilot projects have been undertaken and in feasibility studies to determine the viability of m-Health in the HIV/AIDS sector. The Text-to-Change project uses text messaging in behavior change initiatives. It is a project by a Dutch non-profit organization partnered with the Ugandan AIDS Information Centre and uses text messaging in behavior change initiatives (Phippard, 2012; Chib et al., 2013)

Mobile technology thus has the potential to open up new spaces for more interactive forms of health communication that would be impossible with the traditional, one-way channels of print and broadcast media, although it can also work alongside more traditional one-way health information dissemination channels (Phippard, 2012, p. 131)

6.6 Challenges for Scale and Sustainability

Analysis of the challenges and complexities experienced by m-Health projects and organizations as outlined in academic and grey literature (Table 6.1 below), suggests a number of challenges and constraints. The most pertinent challenges and constraints include: access barriers and divides including gender, language, and literacy; privacy and security; participation of end users and

Table 6.1 Overview of M-Health solutions in HIV/AIDS interventions in Africa

Author(s)	Article Type	Project type	Study period	Study Country	Some reasons for scale up/sustainability or otherwise; Recommendations
Chib et al. 2013	Research report (Text-to-Change)	Pilot Study	2009	Uganda	Sociocultural; organizational – low participant response; non-ownership of phones, gender (women's low literacy), partnering with only one telecommunication provider and therefore subscribers of other providers automatically left out.
Mitchell et al. 2011	Research report	Pilot Study		Uganda	Low access and ownership of mobile phones
Tomlinson et al. 2013	Essay	Not stated	Not stated	Not stated	Text messages, for example, should work under a set of parameters: - when there is follow up - when message is personally tailored - When the frequency, wording, and content are highly relevant.
Haberer et al. 2010	Research report	Pilot study	July-August 2009	Uganda	Individual level collection of healthcare data presents challenges (lack of desire to respond to Interactive Voice Response calls and SMS texts; low prevalence of mobile phones and at all time, etc.). Phone sharing with others must be considered when implementing technology in resource-limited settings and costs.
Bärnighausen et al. 2011	Review article	Not stated	Not stated	Sub-Sahara Africa	Improved adherence to antiretroviral treatment through SMS texts could be short-term and might not persist over time.
Leon et al. 2012	Research article	Pilot project	August-Nov 2011	South Africa	It is uncertain that the potential benefits of m-Health for community-based health services would be retained

Table 6.1 Continued

Author(s)	Article Type	Project type	Study period	Study Country	Some reasons for scale up/sustainability or otherwise; Recommendations
					with immediate large-scale implementation in a health system with a weak ICT environment and limited implementation capacity.
Catalani et al. 2013	Review article	Not stated	Not stated	Not stated	Development of a road map for m-Health implementation science (to improve the uptake, implementation, and translation of actionable research findings into real world programmes) is warranted.
Jackson et al. 2013	Conference abstract	Project description	Ongoing since past 6 years	South Africa	Achieving sustained impact on health outcomes with maternal and child health (MCH) interventions will not be possible unless the gap is bridged between small scale efficacy studies and large scale interventions.
Cell-Life's Mama SMS	Not stated	Project description	Ongoing	South Africa	More support is needed to strengthen prevention of mother-to-child transmission of HIV (PMTCT) programmes, encourage HIV testing and treatment, and avoid loss to follow-up.
Tamrat & Kachnowski 2012	Review article	Not stated	Not stated	World, incl. some African countries	The policy infrastructure for funding, coordinating and guiding the scalable and sustainable adoption of prenatal and neonatal m-Health services and operations require more evaluation of ongoing programmes.
Lemaire 2011	Review article	Not stated	Not stated	Sub-Sahara African countries	A major obstacle to improving and scaling the implementation of m-Health initiatives is the lack of

Table 6.1 Continued

Author(s)	Article Type	Project type	Study period	Study Country	Some reasons for scale up/sustainability or otherwise; Recommendations
					monitoring and evaluation (M&E) and use of meaningful, consistent indicators and rigorous evaluation methods.
Chang et al. 2011	Research report	Pilot	26-month period	Uganda	Key challenges identified included variable patient phone access, privacy concerns, and phone maintenance.
Curioso & Mechael 2010	Article	Not stated	Not stated	Global South	Partnerships between institutions and organizations should be encouraged to support appropriate design, testing, and evaluation of inexpensive and user-friendly devices for usage mainly by patients and health workers.
Heerden et al 2012	Article	Not stated	Not stated	Not stated	m-Health programmes need to move towards sustainability, should take participatory approach and focus on health, not on the technology.
Siedner et al. 2012	Research article	Pilot	June-Sept 2011	Uganda	Network failures are rampant and network reliability across large catchment areas can be a challenge in resource-limited settings.
Cummings 2011	Review article	Not stated	Not stated	Africa	The combination of indigenous languages and illiteracy problems as well as battery charging, privacy and security of healthcare information are major issues to watch in order to sustain a project.
Phippard 2012	Master's thesis	Research study	14 months	South Africa	Key challenges must be addressed, such as sensitivity and adaptation to local needs and socio-cultural contexts,

Table 6.1 Continued

Author(s)	Article Type	Project type	Study period	Study Country	Some reasons for scale up/sustainability or otherwise; Recommendations
					technical and human resource limitations, sustainable funding, roles of governments and private sector, access barriers and divides, including gender, language and literacy, privacy and security, and participation of end users and communities.
Chaiyachati et al. 2013	Research article	Pilot study	2010-2011	South Africa (TB)	Technical challenges, motivations of health care workers within the context of their workflow limitations and improved technology must be addressed for closer, real-time performance monitoring in order to create a scalable intervention that is more likely to improve awareness of adverse events.
Littman-Quinn et al. 2011	Conference paper	Feasibility study	8 months	Botswana	Limited resources, there is need for expansion of the collaborative Relationships between public and private stakeholders, clinicians, researchers, and educators involved with m-Health in Botswana.
Littman-Quinn et al. 2011	Conference paper	Pilot study	2 months	Botswana	The projects are funded, implemented, and assessed by independentpartnerships which have created a m-Health landscape with co-existing, overlapping, and sometimes conflicting sets of hardware, software, and personnel. As m-Health matures, however, there will be need to scale-up using a leaner assortment of tools and centralized administration.

Table 6.1 Continued

Author(s)	Article Type	Project type	Study period	Study Country	Some reasons for scale up/sustainability or otherwise; Recommendations
Thirumurthy & Lester 2012	Research article	Project evaluation	Not stated	Kenya	Need for extensive research on content, frequency and type of telephone communication (e.g. text or voice; one-way or two-way) that can increase m-Health benefits. More research is needed to determine if m-Health interventions can produce enduring behavioural changes that do not wane over time, a whether their effectiveness depends on the populations & medical conditions targeted.
McQueen et al. 2012	Position paper	Project description	Not stated	Worldwide, also African nations	Key factors for m-Health success and sustainability are full government participation and leadership, a strong public-private partnership, and proper coordination.
van Beek 2012	Master's thesis	Research study	Couple of months	South Africa	Obstacles in m-Health: stigma, sexual behaviour and gender inequality, poverty, need for direct contact, and failing technologies.
Salomonsson 2010	Master's thesis	Research study	Couple of months	Uganda	Need to consider youth non access of phones, their non-participation or exclusion in communication process.
Crankshaw et al. 2010	Research article	Pilot study	October-Dec 2007	South Africa	Sharing and privacy, willingness for clinicians to contact by phone, and changes in mobile phone ownership due to theft, loss and/ or damage could all pose challenges to sustainability.
Marshall et al. 2013	Research article	Pilot study	28 weeks	Uganda	Rapid advances in mobile technology are also posing Unique challenges. M-Health projects suffer the same

Table 6.1 Continued

Author(s)	Article Type	Project type	Study period	Study Country	Some reasons for scale up/sustainability or otherwise; Recommendations
					problems that plague other IT projects, such as unrealistic expectations, improper or abstract guidance from management, unforeseen software challenges, and time and budget over runs
Kunutsor et al 2010	Research article	Pilot study	Not	Uganda	The inability to read text messages because of illiteracy and language barriers
Pérez & Pla	Research article	Pilot study	May 2012	South Africa	Some challenges were illiteracy, internet and electricity supply

communities; existing health infrastructure and human resource limitations; sustainable funding; roles of government and the private sector; technical limitations and operational issues. These are expressed within the individual, organizational/institutional, community, and technical/technological levels.

6.6.1 Individual Level Barriers

A key challenge in m-Health is poor access of phones. Access to phones is still a problem in Africa despite the high penetration of the devices in the continent. For example, Table 6.2 below reflects a study that I carried out at an urban hospital in Ghana in 2010-2011. Of the 584 patients on anti-retroviral drugs (ARVs), only 150 (25.8 per cent) used mobile telephones. At a rural hospital with an HIV unit, of the 170 female HIV/AIDS patients, 45 (26.5 per cent) had mobile phones, 123 (72.4 per cent) did not, and two (1.1 per cent) did not provide this information.[7]

In a study in Uganda (Haberer et al, 2010), most of the participants no longer had their own phones or had non-functional phones during the study. Not everyone has a mobile phone, and most people cannot have a sharing of phone arrangements with relatives or friends. Many HIV-positive pregnant women on treatment do not have cell phones or are illiterate. Research has

[7]The hospital's list of male patients was not yet fully compiled at the time of my fieldwork.

Table 6.2 Mobile phone users among HIV patients on antiretroviral therapy (N=584)

Males	148
Females	436
Employed	525
Unemployed	63
Have mobile phones	150
No mobile phones	434
Over 40 years	254
Under 40 years	330

shown that women in sub-Saharan Africa are 23% less likely to own a mobile phone than men (Leon et al., 2012). These vulnerable groups will continue to be left out of m-Health interventions, and hamper scale and sustainability of interventions. Non-literacy and inability to read SMS messages is another constraint.

Again, many AIDS patients at the developed stage of their ailments become weak and unable to walk or work to earn a living for themselves let alone the household (Crentsil, 2007; Ntonzi, 1997). Cost will be worse, and their lethargy means that some other helper would be required to receive and make calls (if costs can be managed), which concerns issues of privacy and confidentiality. Although research indicates that many HIV patients in Africa would be comfortable receiving HIV-related information through their mobile telephones (Mitchell et al., 2011), privacy and confidentiality are challenges due to stigma and shame associated with being HIV positive. Busy signal and network congestion, patients' intentional call rejection or refusal to respond to SMS texts are some of the challenges.

6.6.2 Organisation/Institutional Controls

In Africa, obstacles for scale up and sustainability include key challenges in accountability and lack of coordination. During my study in 2011 in Ghana, the process of using mobile phones in HIV initiatives did not seem to be systematically organized, and policy infrastructure coordinating and guiding the sustainable adoption of m-Health services was under-developed. Research also indicates lack of commitment by governments and their harsh policies against mobile phone companies as some challenges.

Also, there is often an anecdotal nature of project reporting to date, resulting in a dearth of critical discussions and analyses of the actual implementation challenges inhibiting the successful deployment of these interventions (Phippard, 2012, p. 164–5).

It also seems that many phone-based health communication, information, and developmental services are consigned to be small-scale pilot projects only as long as they are funded by international organizations (Tenhunen, 2013, p. 5). The problem still remains of how various stakeholders in the mobile phone, service providers, government agencies, NGOs, and other industries can co-operate to provide applications in affordable and sustainable ways (Tenhunen 2013).

6.6.3 Community Action

Although the dynamics involved in m-Health design and implementation vary according to specific contexts, successful deployment of any health or development initiative depends explicitly on sensitivity and adaptation to local needs and socio-cultural contexts. As the analyses in this paper indicate, there were instances of low participation response (see Chib et al., 2013). In many rural areas, it is uncertain that potential benefits of m-Health for community-based health services can be retained with large-scale implementation in a health system with weak infrastructures and ICT environment (Leon et al., 2012)

Mostly, outside solutions with limited local adoption is another problem. Most programs usually target only individuals or small groups, which make scale up and sustainability difficult to achieve. Pervasive AIDS-related stigmas that permeate some segments of the society as in Ghana (Crentsil, 2007) and in South Africa (Kalichman & Simbayi, 2004, p. 572) may also pose challenges to implementing HIV interventions.

6.6.4 Financial/Technological Constraints

M-Health interventions are constrained by the technological functions of the mobile phones and networks upon which they are based. Although the technology evolves rapidly, we can identify a number of basic technological limitations that bear upon m-Health communication initiatives. The underlying infrastructure—though not nearly as limited or expensive to deploy as some other ICTs like fixed line telephony or cable Internet—is not always universal. Patchy network coverage, service fluctuations, bandwidth limitations, and otherwise unreliable connectivity invariably constrains m-Health interventions, especially in the most rural or remote areas (Phippard, 2012; Siedner, et al., 2012).

6.7 Suggestions for Achieving Scale and Sustainability In M-Health Solutions for HIV/AIDS in Africa

Scalability and sustainability are two critical factors in achieving long-term success and meaningful impact of m-Health interventions in AIDS and development contexts (Phippard, 2012). Although failure rates specific to m-Health are unknown, it is estimated that 30–70 per cent of all health IT projects fail (Sanner, Roland, & Braa, 2012). Lemaire (2011) outlines a number of successful cases in which m-Health projects have been scaled up or are in the process of being scaled up. These include Child Count, mPedigree, mTRAC, Text-to-Change, SMS for Life, and Txt Alert. The failure of most projects to be scaled up or sustained does not mean that things cannot be improved. It has been suggested that in some circumstances, health promotion effects will be sustained without the need for on-going intervention, and efforts to sustain programs are usually not warranted (Swerissen & Crisp, 2004). But this is, perhaps, not with mobile technology in the HIV/AIDS sector. As Haberer et al. (2010) have noted, for adherence to ART, sustainability is critical for double viral suppression to prevent AIDS-related morbidity and mortality.

Attempts to deliver health and other developmental services via mobile phone applications with the help of service providers have been successful in Ghana, South Africa, Uganda and other parts of Africa. Some of these pilot cases have proven that it is possible to use mobile technology to deliver information on health issues.

6.7.1 Individual Level

Much is known about the effectiveness and sustainability of health promotion interventions that are aimed at individual behavior change knowledge, attitudes and beliefs that are the precursors of behavior change (Swerissen & Crisp, 2004, p. 124). These could be adopted in m-Health programs. Interactive and individually tailored interaction programs for behavioral health risks lead to higher levels of sustained behavior change. Mobile companies could coordinate with funders to provide low-prized phones to rural dwellers.

A key potential is how language in SMS messaging can be developed for those who cannot read and understand English. There is a need for services and software in local languages and dialects. Graphics on HIV/AIDS and texts in local languages could evolve to help in the campaign against HIV; the barrier of illiteracy can be addressed through the help that illiterate people may receive from others to send text messages or punch in numbers to make a phone call. Another area with potential for HIV/AIDS information and education is SMS messaging since it reaches people with whose phones are switched off as soon as they are switched on. To address privacy and confidentiality, codes could be developed for only the patient to access the particular information.

6.7.2 Organisational/Institutional level

International non-governmental organizations and the United Nations have been supporting the implementation of m-Health and other HIV intervention programs in cooperation with government agencies. The private enterprise can offer more in collaboration with the public sector. For example, as Tomlinson et al. (2013) have pointed out, emerging internet companies such as Google, Yahoo, and Facebook provide informative case studies of data collection. These and other private enterprises have been outstanding in this direction, and m-Health needs to utilize their platforms and methods to optimize and sustain personal and public health.

Successful large-scale m-Health systems will require investments by African states, NGOs, private companies, and others. Again, mobile industries can push for m-Health scale up, only after research evidence for perceived advantages. The success stories outlined by Lemaire (2011) seem to have been the result of partnering with a mobile technology company, international funders and/or government institutions. Governments also need to consider how to address electricity and energy issues.

6.7.3 Community Action

As Heerden, Tomlinson, & Swartz, (2012) have suggested, m-Health programs need to move towards sustainability but should take participatory approach and focus on health, not on the technology. The belief that HIV/AIDS is caused by spirits and the supernatural demonstrate a misinformation about AIDS and may endorse social sanctions stigmatizing patients, and these should be borne in mind in any m-Health solutions in communities.

The environmental and social consequences for health-related action need to be directly addressed by working with individuals as well as small groups and the community in order to make sustainability easy to achieve. For instance, mobile technology tends to amplify existing structures, but by helping to blur cultural boundaries it also creates spaces for agency and critical discourses: the same technology can empower women, especially

HIV-positive ones, as well as amplify gender disparities and vulnerabilities (Tenhunen 2013). Hence, for example, interventions for pregnant HIV-positive women should also target their partners, households and even the whole community.

6.7.4 Financial/Technological Resources

The health system can also negotiate financial input with mobile phone service providers, government bureaus, international and local donors, as well as with mobile phone users in a way that will be beneficial to all.

Mobile health cannot overcome all of the factors that can hinder adherence to treatment: medication side-effects, high pill burdens, high dosing frequency, lack of trust in the health-care provider, lack of time and lack of money, to name a few. However, by applying m-Health interventions to as many individuals as possible and reserving more resource-intensive interventions for the patients most in need of them, we could increase the overall cost-effectiveness of adherence interventions. Even if m-Health interventions are cost-effective, who will pay for them remains unclear. In settings with a complex mix of public and private health-care providers, everyone has different priorities, funding mechanisms and funding cycles, and such differences can affect the uptake of innovative interventions. Implementation science is an emerging field to address such issues in health promotion and it should be encouraged. Patients' willingness to pay for inexpensive periodic text messages is also worth exploring (Thirumurthy & Lester, 2012)

An m-Health robustness would be its ability to reliably scale up despite variability in quality and coverage of wireless communication networks and limited access to stable power supply (Sanner et al., 2012, p 156). Still, mobile phone service providers should do more to improve the quality of their networks. That many users try to stay connected by finding places with better network coverage or using car batteries to recharge their phone batteries shows how ubiquitous the device is, and how mobile phone companies can capitalize on that to deliver better services for profits.

6.8 Conclusion: Future Directions for Research

Mobile health programs could be effective and scaled up nationally and expanded to include all, with the right attitudes and approach. Mobile health research provides quite impressive evidence for the validity of m-Health approach. However, there is still a need to reflect and assess progress in the context of scale up and sustainability issues and to examine challenges to m-Health initiatives in order to understand the interplay between context and factors promoting the programs and those that inhibit them. Both practitioners and researchers have argued for greater dialogue and attention to these potential barriers and constraints, and have articulated the need for theories and frameworks that transform anecdotal evidence of experienced challenges into enabling and actionable information to guide future project development (m-Health Working Group, 2012).

The lack of sustainability and scalability has been a serious problem with m-Health pilots in low-resource contexts (Curioso & Mechael, 2010; Lemaire, 2011). Analysis of the projects implemented by organizations such as Cell-Life and other pilots and application reveals significant promise and diverse range of opportunities for m-Health (and development) in HIV/AIDS contexts in sub-Saharan Africa. However, this analysis also reveals a number of particular challenges, issues, and complexities that must be addressed in order to develop effective m-Health programs, as well as potential pitfalls and barriers that might inhibit sustainability and scalability. As Luk and colleagues (2009) have suggested, implementation and problems at the institutional, societal, and individual levels, such as lack of logistics, poverty and other socio-economic crises must be addressed.

Mobile phones' benefits, such as the ability to disseminate health information more quickly, thus saving time and reducing shame and stigma, also call forth the capacity of the devices to blur spatial boundaries by acting as a catalyst for reorganization and new interpretations of culturally constructed spheres and boundaries (Tenhunen, 2013).

More extensive research is needed on the content, frequency and type of telephone communication (e.g. text or voice; one-way or two-way) and how it can increase the benefits derived from m-Health applications. More research is also needed to determine if m-Health interventions can produce enduring behavioral changes that do not wane over time, and whether their effectiveness depends on the populations and medical conditions targeted. To shed light on these issues, longer-term studies in many different settings are required (Thirumurthy & Lester, 2012).

Mobile health interventions could be applied to a very broad range of health-related behaviors, although what works in one context may not necessarily work in another. For example, text message reminders may readily help patients to adhere to ART or get health-care providers to follow a malaria treatment protocol, but they may be of no benefit in connection with other health-related behaviors. Even so, I am in agreement with Bärnighausen et al., (2011) that there is need for more research to examine the effect of context and specific features of intervention content on effectiveness. Rigorous evaluations of small-and large-scale interventions could reveal the extent to which m-Health can provide cost-effective solutions to public health challenges. Future work needs to assess interventions targeting and selection of interventions based on individual and community participation, organizational and institutional capacities, as well as the commitments of mobile phone providers and other stakeholders.

7

Mobile ICT and Education Delivery

Nana Kofi Annan
Wisconsin University Ghana
nkan@outlook.com

George Orleans Ofori-Dwumfou
Methodist University Ghana
g_ofori@yahoo.com

Benjamin Kwofie
CMI, Aalborg University
Copenhagen
bkwo@cmi.aau.dk

7.1 Introduction

Appreciating the fact that the application of various forms of information and communication technology (ICT) in education obviously has great potential to facilitate education from primary to tertiary, this chapter takes a specific look at the use of mobile-ICT in education in developing countries specifically in Africa. The term *mobile ICT* has become an Information Technology (IT) jargon and is used in many books today because of the proliferation of portable mobile computing devices coupled with mobile telecommunication. It is believed that one of the ways out for Africa in advancing education delivery is to take full advantage of the numerous opportunities offered by mobile computing and communication technologies.

The focus of discussion in this chapter is centered on mobile learning (m-learning) and its usage in education institutions in Ghana. Although the root of m-learning can be traced to the 1960s, it is undoubtedly true that it is recently that this phenomenon of using mobile devices for teaching or learning started gaining popularity in formal education circles. The growing interest of students and teachers in m-learning is attributed to the advancement in telecommunication, mobile devices[1] and mobile apps[2]. Even though Africa is in its initial stage of harnessing the full potential of m-learning, it is worth noting that there are several m-learning projects and initiatives across its length and breadth with appreciable success stories. A few of these m-learning cases are discussed in this chapter.

7.2 Brief Overview of M-Learning in Africa

Africa can be considered as having the highest mobile learning growth rate in the world with Nigeria being one of the main drivers of the growth rate witnessed across the continent. According to the New Ambient Insight Report, the growth rate is 38.9 percent, which is the highest in the world. It can be said that mobile devices are now the primary computing devices used by many people in Africa. Accessing the web on an Internet-enabled feature phone or a smartphone is often a user's first Internet experience.

For many people in the Africa region, mobile learning is their primary learning technology and they may never be exposed to other learning products. According to *"The 2012-2017 Africa Mobile Learning Market"(Business Day, 2013),* Nigeria is among fourteen countries that are regional drivers of the growth in the mobile learning market in Africa. Others include Algeria, Angola, Ghana, Kenya, Mozambique, Rwanda, Senegal, South Africa, Tanzania, Tunisia, Uganda, Zambia, and Zimbabwe.

There are several organizations which are collaborating to support mobile learning research and initiatives in Africa. Notably among these organizations are UNESCO, DANIDA, NEPAD and Becta. In addition, eLearning Africa Conference, M4D (Mobile for Development) conference among others are also contributing tremendously in promoting the use of mobile learning platforms in schools in Africa. Through the efforts of some organizations and other individuals Africa can enumerate several mobile learning pilot projects and initiatives. The primary objective of all these m-learning projects is to seek ways of making education accessible to all, either as supplementary, complementary or comprehensive learning platform for teaching and learning that engages the interest of the user (Traxler & Kukulska-Hulme, 2005), (Traxler & Leach, 2006). It is also anticipated that by promoting the use of

---

[1]Mobile Devices: This generally refers to handheld computers which includes smartphones, tablets, digital readers, Personal Digital Assistant (PDA) and any type of computer which is portable and can be carried around in one's hand with ease. The mobility of these devices are empowered by wireless network and communication.

[2]Mobile Apps: Is an IT jargon which means mobile application software.

mobile computing and communication technologies in education, it will help in bridging the educational gap which exists between Africa and the developed economies.

Presently, some educational institutions including but not limited to university of Pretoria in South Africa (Ally, 2009; Keegan, 2005), Makerere University in Uganda (Kajumbula, 2006), and University of Ibadan in Nigeria, Keta Secondary School and Central University College in Ghana among others, are using m-learning on pilot basis and more are showing interest in the whole idea of m-learning. For example, the faculty of health science in university of Cape town saw a need to communicate with students in ways not accommodated by current online methods by introducing an m-learning pilot project in January 2005 (Masters, 2005), the BridgeIT initiative in Tanzania, also provides teachers with the access to digital video content for on-demand screening in class via mobile technologies, and Nokia's Mobile Mathematics (MoMath) project in South Africa (Isaacs, 2012a, 2012b). This has been possible because most institutions and governments with the help of donor agencies and telecommunications companies are putting in place, logistics and frameworks that could advance learning using mobile technologies (Frohberg, 2006).

In South Africa, the Square Kilometer Array (SKA), in collaboration with Intel South Africa has launched several community projects in the tiny Karoo town of Carnarvon as a way of propelling the surrounding communities into the information age (IT News Africa, 2013). Intel is working with the SKA and the Department of Science and Technology to supply computers, educational materials, teacher training and internet access to the Carnarvon community center and five schools in the three towns closest to the main SKA site – Carnarvon, Williston and Van Wyksvlei. These schools form part of the SKA e-Schools project, recently Intel launched ten classrooms and community center equipped with Class-Mate computers, teacher laptops, content, servers and printers in mobile trolleys from Smart-labs, and free curriculum-based learning solutions through Intel's recently launched Explore and Learn Education solution.

Each of the schools and community center received a server for hosting purposes as well as internet access, while Carnarvon High School received a mobile science lab from Smart-Labs. This initiative was to provide the people with the best opportunities in getting good education.

7.3 The Emergence of Mobile Learning

The notion of using mobile device for teaching and learning is complexly rooted in technology portability and mobility, human mobility, pedagogical needs, ubiquity of technology, wireless network and communication. This propelled by the evolution of information and communication technologies which comes along with great opportunities for teaching and learning allowing people to access educational resources from anywhere at any-time via mobile broadband network. This new approach in acquiring education is known as mobile learning. It is a rather fast spreading form of learning in this era of high diffusion of mobile devices such as tablet PC and smart phones among different groups of people – students, teachers, workers, farmers, market women, children, young people, and adults including illiterates etc.

Technological development of hardware and software in mobile computing and communication is changing the way people think, act, relate to others and interact locally and internationally. (Fuentelsaz, Maícas, & Polo, 2008). Even though the primary purpose of mobile phones were mainly for wireless communication, its explorative use has given brought about a new ways of using it for educational purposes. Thus it was realized that mobile phones have the potential for teaching and learning apart from the core communication function. (Huang, Chiu, Liu, & Chen, 2011). Designers and developers being aware of this have made efforts to improve the computing power, functions, battery strength, screen size and weight among others to meet user needs which includes mobile learning.

Mobile ICT is enjoying rapid adoption and diffusion among all age groups in every parts of the world with an overwhelming use in Africa as compared to other technologies. Particularly most students in tertiary schools have different kinds of mobile devices that enables them to connect to friends and information resources via mobile internet. The call for the use of these devices in education is partly associated with its portability, availability, interoperability, functionalities and educational usefulness.

For example, in Africa, massive deployment of mobile telecommunication and proliferation of mobile devices are gradually moving m-learning from research-led pilot projects to everyday activity where mobile devices are becoming personal tools, aiding people to learn anywhere and anytime either formal or informal (Kukulska-Hulme, Traxler, & Pettit, 2007). Unlike traditional face-to-face classroom style of education, m-learning has numerous advantages in helping less privileged learners who for one reason or the other could not access formal education. Considering the fact that a great number of learners in developing countries are disenfranchised from accessing formal education because of lack of infrastructure and other facilities, mobile learning makes it possible for such people to have access to education without much difficulty which hitherto was not possible.

Using mobile devices for teaching and learning creates opportunities for seamless learning experience which strengthens Technology Enhance Learning (TEL) by way of augmenting physical space, leveraging topological space, aggregating coherent across all students (Jeremy Roschelle & Pea, 2002). According to (Chan et al., 2006), the space of social-cultural development aided by ubiquitous devices is fast affecting and shifting the way students learn within and outside school environment, this brings the issue of m-learning to the fore. M-learning is necessitated by diverse technological and human factors, which includes (1) ubiquitous access to mobile, connected, and personal handhelds (2) the relentless pace of technological development in one-to-one computing, and (3) the evolution of new innovative uses of mobile computing devices. This supported by learning theories is influencing the nature, process and the outcome of learning and has opened a new door for teaching and learning, characterized by seamless learning space which allow learners to learn anywhere anytime just-in-time by seamlessly switching between different contexts such as formal and informal, individual and social learning with the opportunity of extending the social spaces in which learners interact with each other.

7.4 Mobile Learning Projects and Initiatives in Ghana

Five tertiary institutions including the University of Cape Coast, Centre for Continuing Education have been chosen by the Commonwealth to serve as pilot Centre for E-Teaching. The remaining centers are sited in Malaysia, Maldives, South Africa and New Zealand. The project which is initiated by the Commonwealth Education Trust begun in South Africa and Malaysia in 2013 (GBC, 2013).

The innovativeness of mobile phones has found a place in the educational sector. For example, the Ghana government has introduced the Computer School Selection Placement System (CSSPS), which enabled the admission of students into the second-cycle educational institutions through a computerized selection process in 2005 (Gyepi-Garbrah, 2012). Generally, things changed with the computerization of the process opening up for better scrutiny, coordination and access. Now, students can send SMS to find out about their school placements (Essegbey & Frempong, 2011). The West

Africa examination council has introduced an online service which enables students to make enquiries and check for their exams results on their mobile phones ("WAECDIRECT ONLINE - RESULT CHECKER," n.d.). Students in universities and some other tertiary educational institutions use the mobile phone to access relevant information such as time table, transcript, registration, examination results etc.

There are some other m-learning activities in Ghana which is worth mentioning. The first among them is the CocoaLink; an m-learning project initiated by Hershey Company, World Cocoa Foundation, World Education, Ghana Cocoa Board (COCOBOD) and their technology partner DreamOval ltd. ("News," n.d.). The project utilize the use of mobile phone technology to educate cocoa farmers on performance-enhancing and marketing information on cocoa farming. The project was piloted with cocoa farmers in the western region of Ghana. The farmers received information on best farm practices, child labor, health, crop disease prevention, post-harvest production and crop marketing. The others are m-learning services provided by telecom operators – Vodafone Ghana ("Vodafone Ghana?" About eLearning @ Vodafone," n.d.) and MTN Ghana. ("Forum Solutions - Online Conferencing Solution, Teleconferencing Applications For Education, Religion, Governance, Politics, Business, Media, Others," n.d.).

Six schools in Ghana were privileged to be selected for the first phase of the New Partnership for African Development (NEPAD) e-school initiative (IT News Africa, 2012). The purpose of the initiative is to provide young people with the skills and knowledge they need to participate in the global economy. The selected schools were Wa Senior High School, Ola Girls Senior High School, Walewale Senior High School, Acherensua Senior High School, St Augustine's Senior High School and Akomadan Senior High School. The project was piloted using a Public-Private Partnership (PPP) model which was incorporated into the e-school initiative. See Figure 6.1

7.4.1 IREAD Ghana Project

The iREAD (Impact on Reading of E-Readers and Digital content) Ghana Study was a pilot study which was conducted from October 2010 to July 2011. It was categorized as a Global Development Alliance (GDA) program between the United States Agency for International Development (USAID) and Worldreader, a non-profit organization (WorldReader, 2012).

The purpose of the pilot study was to give Ghana public school students access to books through e-reader technology, which is an electronic device

Figure 7.1 Secondary school students first experience of m-learning in ghana

that can house thousands of books. The iREAD program aligns with USAID's Strategic Objective 8 (SO8) to "Improve the Quality of and Access to Basic Education." iREAD supports SO8 Intermediate Result 2 to "Improved Quality of Education," through (1) Increased access to a number and variety of books and other supplementary reading materials read by the participants of the study (2) Improved student performance on standardized tests of reading, writing, and English proficiency among study participants (3) Reduced waiting periods in classrooms for classroom material and (4) Reduced net cost of production, translation, and distribution of supplemental reading material.

The principal aim of the iREAD Ghana Study was to collect data to determine whether iREAD interventions had any effect on access to reading materials, student reading performance, and overall academic environment.

7.4.2 The CUC M-Learning Initiative

Central University College (CUC) has a remarkable m-learning experience as compared to other universities in Ghana. As part of CUC's objective to educate people to become useful in society, it also appreciates the fact that society is dynamic and changing fast hence, the need to adopt and adapt to new ways of framing and presenting education especially to the 21st century student. The university takes cognizance of the rapid growth in ICT and comes to the conclusion that today's learner must be trained with tomorrow's technology to make them employable tomorrow. CUC has student population of over 10,000. In 2009, CUC decided to introduce into its educational framework, an educational technology which offers students and faculty the opportunity to engage in teaching and learning using their handheld mobile computing devices anywhere anytime. The whole idea as conceived by the academic board and board of regent was to explore and utilize ICT for the betterment of university education. Knowing that ICTs can be used in different ways in education, and appreciating the mobility of students and high penetration of handheld mobile device, the university's board was particular in getting a teaching and learning system which utilizes mobile computing and communication technology, capable of supporting academic activities outside - bricks and mortar. Thus, a ubiquitous platform which allows 24/7 access to academic resources and without geographic limitations.

In view of the above, mobile learning was introduced at Central University College (CUC) in Ghana to augment the traditional face-to-face classroom education to facilitate and extend teaching and learning anywhere anytime. The project was initiated in 2010 and it was the first of its kind to be piloted in a university in Ghana. The whole idea was to use mobile ICT in ways which will be meaningful and helpful within the university's educational framework. Some of the antecedents which necessitated the project were mobility of academic activities, increased large class sizes, and technology affordances in education delivery in the 21st century among others. The implementation of the m-learning platform was a pilot project which spanned over two and half years from 2010 to 2012. The purpose of the project was to help faculty in managing paperwork, distractions, perceived anonymity of students while adding flexibility in teaching and learning and more especially to meet diverse teaching and learning styles. The outcome of the project was to help improve performance of faculty, students and administration. The project was implemented using action research which allowed the implementation process to be iterated. By using action research, the researcher and the practitioners had the affordance to learn, relearn and unlearn through the iterative nature of the action process until the desired result was obtained.

CUC engaged the service of AD-CONNECT consulting services an IT company based in Accra, Ghana to provide them with an m-learning platform. The company developed an m-learning platform which enables faculty to develop and publish content from anywhere on any time for students to access with their mobile computing devices at any-time from anywhere and which included features like forum, chatting, SMS, blogging and activity tracker. The anticipation of the project was to transform teaching and learning at CUC to reflect current trends of using educational technologies to facilitate education delivery which is learner centered and able to meet the educational needs of the learner.

Among the urgency of CUC as the first educational institution in Ghana to implement m-learning were but not limited to (1) delivering education in different ways which accommodates diverse learning styles of students to make learning flexible, interesting and enjoyable (2) help faculty to manage large numbers of students in terms of conducting exams and marking scripts. (3) to keep students and faculty daily updated on academic activities through the SMS alert functionality of the m-learning platform such as 'schedule', 'prompt' and 'reminder' (4) to foster collaboration among students and faculty through the social media forum on the m-learning platform for discussions, feedbacks and sharing.

The project as envisaged by the stakeholders was to effectively harness mobile computing and communication technology and incorporate it into the pedagogical framework of CUC to enhance teaching and learning among other things. The project was motivated by (i) rapid advancement and proliferation of m-learning projects worldwide (ii) the unprecedented development of mobile computing and communication technology which makes it possible for it to be used as mediating tool for teaching and learning activities anywhere anytime (iii) the mobility affordance of mobile technologies in the delivery of education (iv) the availability of portable mobile computing devices in the hands of teachers and students – high adoption and diffusion among students (v) the improvement in mobile telecommunication service and infrastructure in Ghana which now has the capacity to support 3G and LTE with high speed data transmission and (vi) competition among mobile telecom operators which make services relatively cheaper.

The whole project was considered as experimental in phases. It started with an initial pilot (1st phase). This lasted for two semesters of three months each with 500 students and 22 lecturers with 44 subjects including English, physician assistant, nursing, economics and accounting. The subjects were picked randomly from level 100 (1st year) to 400 (4th year) with each level having a taste of the new learning technology. The second and third phase followed in full scale where the whole school population of about 10,000 students were enrolled onto the m-learning platform. The trials were quite successful, some of which were (i) software in ability to meet pedagogical needs (ii) poor human computer interaction (iii) lack of constant 3G coverage anywhere anytime (v) additional cost and (vi) most students had clone mobile devices from china which did not connect well to the 3G network at all times.

7.4.3 UGBS MOBILE – A Library Powered by Mobile Devices

The University Ghana Business School has introduced a mobile library project. The project is known as UGBS Mobile ("UGBS MOBILE - A Library Powered by Mobile Devices", n.d). It is a library which is powered by mobile devices as shown in Figure 7.2 The motivation for the project is to provide faculty, researchers, students and general public with multimedia educational content primarily through mobile devices. The project was sponsored by University of Ghana Business School, Pearls Richards Foundation, Samsung West Africa, Craft Concepts Limited, Ghana Multimedia Incubator Centre and Business House JCR 2012/2013.

As part of supporting government's efforts in promoting education delivery through the use of ICT, the UGBS Mobile initiative is to offer educational institutions and students with multimedia educational content on mobile devices thus to deliver educational and information materials electronically through mobile devices. The purpose of the project is (1) to equip students and faculty in Ghanaian tertiary educational institutions with interactive content-driven mobile applications relevant for teaching, learning and research (2) provide students with skills to develop mobile applications of relevance to the Government of Ghana's goals on good governance, health care, agriculture and education and (3) provide training in technology entrepreneurship for students through mobile application development.

The project is in two phases. The first phase is the establishment of the UGBS Mobile Space at the University of Ghana Business School. This consists of seven mobile and computing devices in a room of 15 seating capacity. The focus for the first phase is to provide educational applications and electronic resources to serve about 3,000 students in the University of Ghana Business School. To train an average of 200 students per year in mobile application development, in a bid to enhance the entrepreneurial skills and provide users with practical experience in support of courses taught in the University. In addition it will develop applications relevant to the delivery of government services and informational services in health care, agriculture, education and government. The second phase of the project is to extend lessons in the University of Ghana Business School learnt to other departments in the University of Ghana and other public and private universities in the country.

Figure 7.2 UGBS mobile library

The relevance of the project was in line with the university's goal to develop world class human resources and capabilities to meet national development needs and global challenges through quality teaching, learning, research, and knowledge dissemination.

7.4.4 Danish IT Ghana LITE Model Project

The Ghana Learning Innovations Teams for Education (Ghana LITE) Model Project was an initiative to transform classrooms into active learning environments for pupils and their teachers with the help of trained coaches to explore different ways of learning to maximize their learning experiences and satisfy their learning needs. It sought to provide teachers and pupils access to large quantities of open, digitized, multi-media educational resources to facilitate the teaching and learning (OLE Ghana, n.d).

The project had an embedded video feedback which enabled both teachers and learners to see how well they were creating an active learning environment. Teachers and students were expected to eventually be able to modify existing content and create their own, thus giving them a sense of agency in the learning process as distinct from being passive receptacles of information.

The model project was to test the effectiveness of the Ghana LITE initiative in increasing student academic performance over a stipulated period of time. The model project was carried out in a single basic school focusing on literacy and numeracy for primaries 4, 5 and 6.

The model pilot (including resource development, site selection and preparation as well as classroom coaching) run from October 2011 to December 2012 (14 months). Funding for this project was provided by the Danish IT Society, an association in Denmark of more than 5,600 members who specialize in a wide range of information technologies. Open Learning Exchange (OLE) has enjoyed a long and productive relationship with the Society and its members. Very useful lessons have been learnt from this model project. The Ghana LITE project gave birth to the Ghana Reads Program.

7.4.5 Ghana Teacher-Mate Trials

The Ghana Teacher-Mate Trials primarily sought to assist teachers with hand held devices, specifically Teacher-Mates, and its accompanying Differentiated Learning Systems to cause significant improvements in the literacy levels of pupils at the basic level. It also sought to prove that when pupils and teachers are exposed to quality teaching and learning resources, with very little guidance they can enjoy a much better learning and teaching experience. The trials were carried out in a single basic school with the focus being literacy for primaries 2, 3 and 4.

It was therefore expected that there would be increased levels in pupils' literacy performance over a specified period of time. The production of a school newsletter and other literary artifacts as a physical manifestation of the direct result of the Teacher-Mate intervention was also to be encouraged. It was also envisaged that there would be a positive attitude towards teaching and learning in the school manifested by a possible increase in enrolment, regular as well as punctual attendance to school and increased general academic performance (OLE Ghana, n.d).

7.4.6 Ghana Reads

Open Learning Exchange (OLE) Ghana proposes the Ghana Reads project to demonstrate and document an approach that can accelerate literacy significantly among all of the children of Ghana. This involved developing, deploying and assessing the strengths and weaknesses of an expanded national network of low-cost digital elementary school libraries that is scalable.

The Ghana Reads installed and supported School Bells (School Basic e-Learning Libraries) in twenty rural elementary schools in the Ga West Municipality of the Greater Accra Region. With high quality open education resources (OER) it enables all of the students, guided by their teachers and with the help of expert coaches, to become literate. A Ghana National BeLL was used to update the School BeLLs periodically with new resources. At the same time data on uses, comments and ratings of the effectiveness of the resources in improving students' achievements were collected. The intention is to provide Ghanaian educational resource developers and policy makers with rich data that can help them improve the effectiveness of their learning materials and strategies for achieving universal child literacy.

The BeLL system also sought to encourage teachers and students to adapt the open source content found on the BeLL for local conditions, and for children with special needs because it is flexible for one to create and publish entirely new content. Communities were to be encouraged to use the BeLL to produce and distribute their own local newspaper and to encourage adults, as well as children, to use it to expand their own reading and writing skills.

The School Bell system works off the Internet, as well as on. The BeLL server contains two terabytes of multi-media resources, either OER or licensed for all Ghanaian elementary schools, that can be customized and updated to meet the needs of a particular setting. The server, requiring twelve volts of power and using 60 watts, can be powered locally if needed. The School BeLL system includes a projector, speakers, HD video camera, laser printer, and Wi-Fi supporting a variety of hand-held mobile devices. When the BeLL is not connected to the Internet, it can be updated periodically with a flash drive taken to the nearest Internet Café or hot spot. With the hardware cost of under US$1,000 per BeLL, and assuming a four-year hardware life in a school with as few as 250 students, the cost per student per year for a complete library of high quality free educational resources is approximately one US dollar.

The project was guided by the success story of the 2011 TeacherMate literacy project in Rwanda, from OLE Ghana's 2012 replication of that program and from the current Ghana LITE project. In January 2012, with support from the Danish IT Society and Barnes & Noble, OLE Ghana's LITE project deployed a beta version of the School BeLL at the Katapor School in rural Ghana. This project is advancing abilities to achieve high levels of literacy among young, marginalized children (OLE Ghana, n.d).

7.5 Potentials

There are several factors which are known to be advancing the cause of mobile learning in Africa. Notable among them according to UNESCO report 2012 are the exponential growth of mobile phone subscriptions in the region (Shafika, 2012), which is due to unprecedented development in mobile technologies and low costs for mobile devices and data plans. The liberalization and deregulation of the telecommunications sector has paved way for massive private investment in the mobile telecommunication industry. Systemic failures in the delivery of traditional education are another driving factor in the development of mobile learning in Africa. As is the demand to increase the scope, scale, quality and equity of education paramount to the global challenge of reaching EFA goals. The quest to tackle these issues creates the platform to explore innovative ways for education delivery.

The pervasive access to mobile phones, particularly among the youth, has great potential for expanding learning opportunities to deprived communities which are at risk of exclusion from affordable, good quality education. A related driver is the potential of mobile learning to enable

Open and Distance Learning (ODL), which is becoming an increasingly popular educational option in the rural areas of Africa (Mishra, 2011). Mobile learning is inspired by the new ways in which young people are using mobile phones to communicate and share knowledge. According to (Shah and Jansen, 2011; UNICEF, 2011), there is a potential groundswell of emergent social and educational uses for mobile phones that could drive further mobile learning developments, particularly as mobile technologies advance over the coming years.

7.6 Challenges

According to Annan 2013, there were numerous user concerns which hinder the progress of mobile learning. These issues include but are not limited to computer self-efficacy, perceived ease of use, perceived usefulness, motivation, and job fit, relative advantage, affect, social cultural factors. The first set of factors which were among the key elements of ensuring effective and efficient use of mobile learning are the user's knowledge in the technology he or she is being introduced to, his or her interest in using such technology and the availability of the basic skill that is required to use the technology. For example, lack of skill creates inconveniences to some users, lack of interest caused some of them to distance themselves from the technology and for lack of knowledge, most of them cannot appreciate that the technology is capable of promoting education. Another impediment is basic prerequisite ICT skills. Lack of these skills has a negative effect on effective use mobile devices for teaching or learning purposes. The lack of skill is synonymous to lack of confidence which deter user.

The amount of time needed by a user to perform an activity using the m-learning platform and also the system, service and information quality are very important to the users. In a way, these factors augmented their interest and motivation. It is observed that unnecessary waste of time in using any mobile learning system due to cumbersome procedures and some few hiccups in the quality of the system, service and information tends to have some negative impact on the implementation.

To a large extent, the demography and social cultural background of users affect the use of mobile learning. Key among these factors are gender, age, beliefs, ethics, social status and exposure, attitude, perception, geographic considerations, individual and organizational culture. For example some believed that using mobile devices all over campus and everywhere in the name of teaching and learning is ethically unacceptable to them because they come from a background where they are not used to using such technologies in this way. Some also cannot phantom how students could interact with lecturers and their colleagues 24/7. They perceived that using the technology this way could be intrusive and disruptive. Lack of motivation from leaders in addition to lack of infrastructure can also demoralized people from using the technology.

7.7 Conclusion

It is well known that television, personal computers, laptops, interactive white boards and projectors have been very useful in instruction design and pedagogy delivery (Koszalka, T. A., and Ntloedibe-Kuswani, G. S, 2010), (Anderson, 2006). If these technologies are seen as offering good opportunities for education delivery, then it must be well noted that the amalgamation of mobile communication, mobile devices and mobile applications builds on the drawbacks of these former technologies to provide a better way of making teaching and learning available anywhere anytime, diffusing geographic constraints. The ubiquitous nature of mobile technologies continues to spark interest among education and technology stakeholders in incorporating these technologies into education. Although developing countries are lagging behind the western world in education delivery, it is good to know that the escalating number of mobile education research especially in Africa gives positive indication that there is enough evidence that they can close up the education gap with the use of mobile technologies in education. For example, the growth in the use of portable and ubiquitous computing devices in Ghana has provided a tool to enhance access to teaching and learning resources just-in-time.

From a pedagogical perspective, using mobile devices for teaching or learning offers context of authentic learning resources in the learning activities (Jeng, Y. L., Wu, T. T., Huang, Y. M., Tan, Q., & Yang, S. J. H, 2010). Though, there are different learning philosophies in education psychology, it is possible for mobile communication technologies to be designed and modelled to suit different pedagogic strategies such as computer-supported collaborative learning, cooperative learning, generative learning, action learning, computer assisted instruction, distributed learning among other (Huang, 2009), (Lundin, J., & Magnusson, M, 2003). For example, blending face-to-face classroom education with mobile learning could be very useful in enhancing teaching and learning. The high penetration, adoption and use of mobile technologies in Africa is an indication that the use of mobile technologies can offer great opportunities in extending education to a lot of people who are illiterate.

In universities, mobile learning helps educational institutions to enhance the accessibility, interoperability and reusability of educational resources, and also to improve flexibility and interactivity of learning behaviors at convenient times and places (Murphy, 2006), (Kukulska-Hulme, A., Evans, D. and Traxler, J, 2005).

From the view of society, mobile learning not only extends learning opportunities to the traditionally hard to reach learner communities, for example, dropout teenagers, unemployed learners and learners with learning disabilities, but it also provides a practical alternative to implement informal learning and lifelong learning.

In many parts of the world, mobile learning is becoming a new booming industry sector and provides new business opportunities for merchants (McKinsey, 2012).

8

Dissemination of Climatic Information and Market Driven Extension Services to Smallholder Farmers in Africa Using Mobile Technology: The Case of Esoko Ghana Commodity Index

Patrick Ohemeng Gyaase
CMI, Aalborg University, Copenhagen,
Denmark
pakw@cmi.aau.dk,pkog@cug.edu.gh

Kwadwo Owusu
Department of Geography and Resource Development,
University of Ghana, Legon, Accra
kowusu@ug.edu.gh

8.1 Introduction

Post-independent Ghana has continued to depend on Smallholder agriculture for significant contribution to the economy. Agriculture is dominated by crop production in Ghana, with crops other than cocoa accounting for nearly two-thirds of the agricultural GDP. Agricultural exports, comprising cocoa, oil palm, cotton, rubber, and fruits, account for around 20 per cent of agricultural GDP (Nankani, 2011). It also employs 56% of the labor force according to ISSER (2011). Even though agriculture plays such a tremendous role in the economy of Ghana, it is dominated by smallholder farmers on rain-fed basis (Owusu and Waylen, 2013) with little application of technology.

The modernization of the traditional agriculture through the use of modern technology can therefore play an important role in a country's transformation to a modern economy. For instance, the rapid diversification of agricultural marketing has accelerated growth in agriculture and general economic transformation of many emerging developing countries (Kolavalli, et al., 2012).

Given the central role agriculture is playing in promoting growth and poverty reduction in the Ghanaian economy, the need for an agricultural revolution based on productivity growth and infusion of modern technologies to solve some of the age-long problems affecting the sector is long overdue. The utilization of modern technology in agriculture could be the most efficacious way towards poverty reduction and growth promotion. In addition to the problems of agricultural marketing and dissemination of agricultural information to the mainly rural farmer, climate change is threatening the growth in the agricultural sector in Ghana with changing rainfall patterns as the main cause of concern (Owusu, et al., 2008).

The ability to create awareness and the timely delivery of climatic information would have a major impact on living standards among Ghanaian smallholder farmers. Climate Change affects all countries in the world. Droughts and floods are destroying the crops and harvest of farmers in developing countries, leaving them in a miserable situation with changing rainfall patterns especially in sub-Saharan African (Owusu & Waylen, 2013). Climate Change is one aspect of the explanation of how the livelihood of farmers can be threatened. In Ghana, the climate has changed over the last few decades. Crops are getting destroyed due to periods of extreme heat and heavy rains (Müller-Kuckelberg , 2012).

8.2 Approaches to Extension Service and Agricultural Information Dissemination in Ghana

Agricultural Extension Services in Ghana dates back to the nineteenth century with the aims of providing an avenue for research findings and new farming methods to be communicated to farmers. Developing a medium to exchange information is essential for the integration of research outcomes and marketing information, with farming practices and extension agents utilizing technologies that work. In Ghana, the main providers of extension services and agricultural market information has been the ministry of Agriculture, Department of Extension service and non-governmental organizations (MOFA, 2010).

The traditional subsistence agriculture is gradually been replaced by market-oriented or commercial agriculture due to variety of factors including rapid economic growth in both developing and developed countries, introduction of new technologies, market expansion, market liberalization, increased demand for food, decreasing farming population as result of urbanization, liberalized and open economic policies, bilateral and multilateral economic agreements, developed infrastructure facilities in farming areas and government agricultural policies (Reddy & Ankaiah, 2005).

Improvement in general agricultural production, productivity and sustainability will depend on farmers' willingness and access to new technology. Agricultural extension services play a pivotal role in ensuring that the clientele (farmers) have access to improved and proven technologies and that their concerns and needs are properly addressed by relevant service providers (MOFA, 2010).

Agricultural extension contributes to improving the welfare of farmers and other people living in rural areas as extension advisory services and programs forges to strengthen the farmer's capacity to innovate by providing access to knowledge and information (Mittal, Gandhi, & Tripath, 2010). However, the role of extension today goes beyond technology transfer to facilitation; beyond training to learning, and includes assisting farmer groups to form, dealing with marketing issues, addressing public interest issues in rural areas such as resource conservation, health, monitoring of food security and agricultural production, food safety, nutrition, family education, and youth development and partnering with a broad range of service providers and other agencies (Nankani, 2011).

However, due to weak extension systems, varied and heavy loads of extension staff has contributed to low or non-adoption of new agricultural technologies by farmers resulting in poor farmer access to market for their produce. In many countries low agricultural production has been attributed, among other factors, to poor linkages between Research-Advisory Service-Farmers and to ineffective technology delivery systems, including poor information packaging, inadequate communication systems and poor methodologies (Kalra, Rajiv, & Khanna, 2010).

Several approaches have been tested, and adopted by countries in Africa to improve the dissemination of information to farmers over the years among which are;

i. *Training and Visit (T&V)*: T&V is one of the earlier approaches that focused on transfer of technology using a top-down, one-size-fits-all approach. This approach was introduced after the department of agricultural extension services (DAES) had been organized under the unified extension systems (UES).

ii. *Participatory Approaches*: In participatory approach the role of the extension agent is to facilitate an in-depth situation analysis by the farmers themselves at the onset of their working relation. Farmers are thus empowered to identify the causes of their problems and to identify the most pressing ones; the extension agent then provides technical knowledge and technologies that are useful in addressing the challenges, which may be useful to address the problems identified.
iii. *Farmer Field Schools (FSS)*: The Farmer Field School (FSS) is a kind of participatory method of delivering extension services modelled on adult-learning principles. In Ghana FFS are being used for a variety of activities, including food security, animal husbandry, and soil and water conservation. Selected farmers meet regularly during a cropping season where they are educated by observing the happenings in the field. Group discussions and practical management of the field from land preparation to harvesting are used in the process. Some of the participants could be selected to receive further and advanced training to become farmer trainers.
iv. *The Commodity Approach (CA):* This approach is normally an initiative of quasi-state or private institutions. This approach facilitates the vertical integration of the production system from input supply to the technology adoption and marketing of the produce. Farmers who are usually out-growers of a major cash crop such as cocoa or oil-palm are empowered with training and technology to produce a certain quantity and quality of a crop, animal species or animal product, and sell it to the initiating institution. The sponsoring company thus provides the necessary support in the form of inputs, credit, extension services, quality management and marketing services to the farmers in return for their produce.

Although there is lack of empirical evaluation of the effectiveness of the above approaches in Ghana, the success of any agricultural policy would likely be influenced by agro-ecological climate, availability and prices of inputs, market access, and farm- and farmer-specific variables (MOFA, 2010). The failure of many of these extension approaches to meet their goals effectively, coupled with inadequate personnel and limited budgets for supporting public extension, has led to continuous modification and experimentation with modern innovative approaches such as using the mobile technology for the dissemination of market oriented extension services given the increase access and use of mobile phones in Ghana as well as improved capabilities of the mobile devices to handle varieties of contents.

8.3 The Growth of Mobile Technology and Value Added Mobile Services in Ghana

The phenomenal growth of mobile subscription Ghana has been acknowledged world over with subscriptions exceeding exceeding 100% of the total population and forecasted to reach 24.5 million by 2015 (IE Market Research Corp, 2011). The Subscriber base for mobile voice telephone services stood at 27, 511, 659 as at the end of August, 2013 whilst data subscriptions stood at 10, 564, 180 with data services penetration standing at (NCA, 2013).

One of the most significant mobile initiatives in Ghana which is the core of this chapter is the Esoko Ghana Commodity Index (EGCI), a mobile application innovative platform which is taking the advantage of the extensive mobile coverage of one of the major mobile network operators to supports agricultural marketing in Ghana. The application utilizes both basic mobile phone capabilities as well and data services. Esoko is a private software company established in 2005 based in Mauritius and Ghana. The company focuses on revolutionary mobile solution to transform marketing and information delivery.

This mobile innovation provides antidotes to the shortcoming of the various means of delivering market driven extension services and climatic information, riding on the back of the phenomenal growth in mobile subscription both voice and data to facilitate personalization and around the clock vital information needed by farmers. This innovation has been acclaimed by various international organizations and has received support from well-known technology entrepreneurs.

The Profile of Esoko Mobile Commodity Index Platform

Esoko Ghana is the only owned and operated franchisee of Esoko Networks Limited (the Mauritius based parent company). Esoko Ghana Limited formerly known as Trade Net, was officially launched in January 2007 to meet the information needs of famers and other players in the agricultural value chain. The Esoko Commodity Index, provides smallholder farmers an opportunity to sign up to Esoko platform which delivers weekly adversary packages for them in the form of current market prices, matching bids and offers, weather forecasts, and news and other tips. The application could also use voice messages and live call center could also provide expert agricultural advisory services to compliment the data alerts sent to farmers and other customers with voice.

Figure 8.1 Farmers accessing price information

The company thus provides technology platform and consulting service that helps organizations to profile people and personalize their information to be delivered to them. The company focuses on the agriculture supply chain with the objective of improving the transparency of markets and the operational efficiency of organizations involved in the sector. The Esoko platform therefore utilizes the mobile phones capabilities to provide an unprecedented and disruptive opportunity to change how markets work for both businesses and consumers. Financed privately by two individuals organizations, the International Finance Corporation (IFC) and the Soros Economic Development Fund (SEDF), The company believes that private sector service organizations, which are accountable to clients for their revenues and long-term survival, are an effective means of development as well as providing public sector services most of which are inefficiently delivered as evidenced in the provision of valued added and market oriented marketing information for farmers. Thus the company is pioneering various agricultural solutions on mobile platform which provides an unprecedented opportunity for the modernization of agricultural marketing through technology in African.

In addition to providing organizations a powerful set of tools for the management of their information to take advantage of mobile technology, Esoko collects and provides content such as prices, bids and offers, weather, and agricultural tips to which users can subscribe. It does this not only by pushing out alerts and advisories to the field, but also tracking information from the field. Organizations have the options to determine whether their data be kept private, shared, or use other content that is publicly available from any of the other Esoko countries. The Esoko Ghana Commodity Index (EGCI) provides farmers a platform to negotiate better prices, choose different markets, or time sales better, as well as participating in out grower schemes through Esoko profiling and reputational history. Figure 8.1 shows farmers accessing price information on their mobile phone at a demonstration session of the Esoko Commodity Index platform.

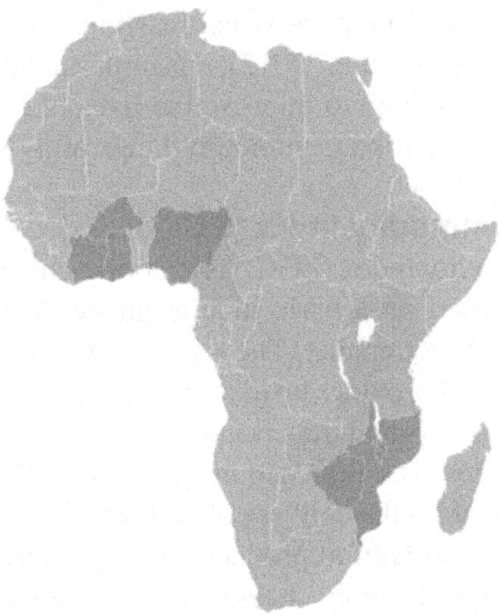

Figure 8.2 Current esoko mobile application operational countries

The objective of this innovative mobile platform is providing smallholder farmers an opportunity and a platform to earn more money. The company's goal is simply to put more money into the hands of farmers by addressing the information asymmetry that exists where farmers are frequently disadvantaged price-takers which often results in the farmers selling at a loss. Esoko Platform thus provides a transparent system that is aimed at reducing time to market for agricultural product, improving profitability; reducing post-harvest loses. This has been done by creating a mobile enabled transaction eco-system that is not necessarily aimed at removing the important role traders play in the food agricultural products supply chain but offer small holder farmers the opportunity to make profitable marketing decision on the sale of their produce with the available price information. (http://www.esokonigeria.com/newsreader.asp?news=1,2013).

Esoko is implementing this innovative mobile platform in partnership with a number of mobile network operation (MNO) across Africa and is partnering a wide range of partners including international organizations, businesses, NGOs, and research institutions throughout the globe. Similar products are being deployed in a number of African countries including: Burkina Faso, Burundi, Cameroon, Cote d'Ivoire, Ghana, Kenya, Madagascar, Malawi, Mozambique, Nigeria, Rwanda, Tanzania, Uganda, Zambia, and Zimbabwe as shown in Figure 8.2.

With mobile market penetration in Africa reaching 63% in 2013, (The World in 2013; ICT Facts and Figures, 2013), it is important that the continent takes advantage of the technology to improve the lives of its people. This is the growth potential that the Esoko Commodity Index aims to harness.

The Esoko Ghana Commodity Index (EGCI)

The Esoko Ghana Commodity Index (EGCI) is a mobile-enabled, cloud-based service which could be used on a basic mobile phone or a computer with SMS and data connection functionality. The EGCI as web-based Agricultural Commodity Index tracks and provides up to date price information of selected agricultural products to farmers and traders among other valuable agricultural information. The application was launched in Ghana in 2010.

The EGCI platform enables the delivery of both wholesale (EGCI-W) and retail (EGCI-R) prices of agricultural products and their locations real-time. The players of the agricultural supply chain who receive this information are thus able to make important marketing decisions. The delivery of weather and planting information also help farmers to make vital planting decision. The users are required to sign on to the platform and once this is done, the players in the agriculture supply chain could request the price information through a designated short code. In Ghana, users are charged a minimal fee of 8 Pesewas per text. The information requested would be delivered in real-time through Short Messaging Service (SMS) (http://www.esoko.com/about/index.php#farmers, 2012).

In the other seven African countries where Esoko has deployed this innovative mobile platform, it has established relationships with private sector businesses through a reseller programme. These private businesses offer Esoko subscriptions and consulting services to local players in the agricultural supply and value chains, offering local solutions to meet local customer demands. In Ghana however, Esoko has implemented its own market information system

Figure 8.3 An enumerator demonstrating esoko application to farmers

Figure 8.4 Sample of messaging interface

(MIS) as a means of testing various deployment methodologies and improve data collection techniques. The lessons learnt from Ghana would thus inform the strategies to improve deployment on other countries.

By this mobile platform, Esoko has been partnered by some donor agencies and international organizations such as USAID, GIZ, FAO and IFAD, for the improvement of inter-regional trade, increase smallholder farmers' livelihoods while improving the flow of information and communication throughout the agricultural value chain. Governments are also taking advantage of the platform to improve their capacity to collect and distribute timely climatic information and market driven extension service to the targeted beneficiaries.

Unlike many donor supported interventions in Africa, the Esoko Ghana Commodity Index (EGCI) is also a market driven application with the ability to provide efficient and effective solutions to agribusinesses and policy makers requiring the collection and management of information from the field.

The traditional method of collecting market data especially in Africa has been to physically visit targeted area and asking questions, sending out questionnaire or making phone call. However with Esoko's innovative mobile platform, data is obtained through scouts who inform the managers at Esoko to setup automatic polls from chosen participants. Simple questions are then sent out via SMS at scheduled times.

Esoko Ghana has also established a network of enumeration agents stationed in market centres. These agents feed Esoko with critical data like prices, offers and industry profiles. Enumerators are well trained and operate throughout the country, coordinated by two managers who set targets, provide on-going support, supervise and monitor data accuracy making prices recorded in the morning on the 'market floor' a part of the index that very day.

The participant's answers are then made available such as the price of particular agricultural products at specific location or planting times for certain crops. This innovative mobile enumeration method used by the Index makes it the most credible information portal for the agricultural sector where access to up to date information is difficult due to difficulty in collecting and distributing relevant information.

The players in the agricultural value chain who receive this information in real time are able to make decisions quicker since prices in other markets are known thereby facilitating profitability. Farmers would also be able to know the prices of their products in different market thus empowering them in the price negotiation process with traders. Traders on the hand would know where to send their stock to make profit. Large buyers especially in cash crop sector such as cocoa could also follow crop yields; weather forecasting and planting activities to enable them predict or plan for harvest yields. The Esoko Platform is therefore a win-win application for everyone in its value chain (Hemen, 2012).

As indicated earlier, the growth of Africa's mobile subscription is among the fastest in the world and defying all anticipated hindrances of adoption such as illiteracy and predominantly rural settlements. Mobile technology is therefore shaping the future of the continents and its economy like never before (The World in 2013; ICT Facts and Figures, 2013).

8.3.1 The Innovative Features of the Esoko Ghana Commodity Index (EGCI)

The (EGCI) platform is the first of its kind in the country. It is a cash market price index composed of data on physical commodities from physical markets located throughout the country. The index not only publishes the prices of individual products in real time upon request but also publishes weekly the average prices and tracks both wholesale and retail prices namely the Esoko Ghana Commodity Index-Retail (EGCI-R) and the Esoko Ghana Commodity Index-Wholesale (EGCI-W). These Indices were developed to provide benchmarks for selected commodity prices in selected markets as a measure of commodity price performance over time. The platform thus solves the problem of the absence of standard warehouse receipt system and a regulatory framework that facilitates the operation of a commodity exchange in Ghana using the knowledge and understanding acquired in the field through the filed agents making the platform appropriate for the market condition.

8.4 Market Price Reporting and Distribution

Esoko mobile platform originally designed to track and distribute marketing information among the players in the agricultural value chain has matured into a feature-rich suite of price data tools. Any of the designated networks or designated members could upload real-time prices from their market locations through the use of SMS syntax, from a cybercafé, or through android smartphone. Different types of prices such as farm gate, wholesale, retail and futures can be captured and the exact location of the prices provided as well as the provision of maximum, minimum and the most quoted prices throughout the country. Multiple price readings from different members on a single commodity or location can be uploaded. Upload could be done at the time intervals of hourly, daily, weekly, or monthly based on the type of product and the choice of the merchant. Price changes are clearly articulated in percentage, large variations are flagged, and tools to contact the original enumerator via SMS provided. Once approved, the prices are available instantly to anyone in the agricultural value chain through web, SMS or Android.

8.5 Bulk SMS Push

Managers of a trade group or registered members of a network can select any number of members within the network, compose freeform SMS messages, and then schedule those messages to be sent immediately, at regular intervals or at a later date. Unique to Esoko, specific users could be targeted with the SMS, using commodity, location and occupation as filters. The Esoko Mobile platform provides interconnectivity among partner mobile operators and can also defers to a number of global SMS gateway providers to ensure that customers are charged the lowest rate possible. Delivery reports are provided indicating the status of messages sent and the platform facilitate the creation and saving of SMS templates, including message and recipients signed up customers.

8.6 Automated SMS Alerts

The Esoko Mobile platform although provides the option for farmers to send a request for price of a specific product to a dedicated short code through SMS , the company identified the need of many of the farmers for regular automated SMS alerts of price changes and other relevant market information. This is done through the setting up of a single user or groups of users whose market

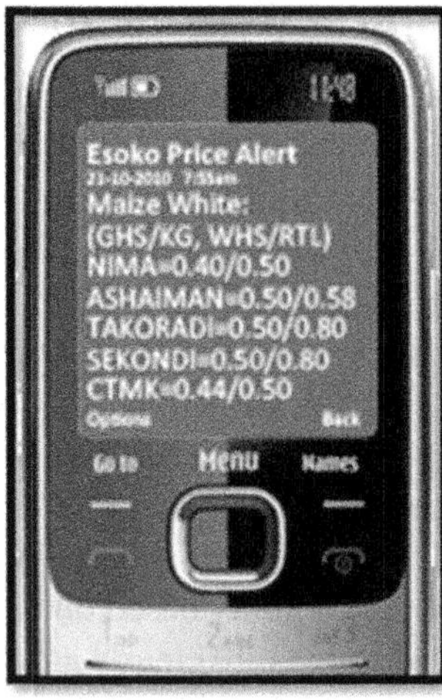

Figure 8.5 Esoko price alert displayed on a mobile phone

SMS are delivered on specified days of the week based on their registered information needs. The Esoko mobile platform currently delivers market price alerts, weather alerts and matching bids and offers alerts. The setting up of such specialized services includes whether it is the recipient who pays for the alert themselves or if the services is being sponsored by an organization or a trade group which is prepared to pay on behalf of the recipients. Automated alerts could also be set up for an entire group such that any addition to membership of the group would result in the added member automatically receiving the relevant alerts and any member removed from the group would cease to receive the alert automatically. A display of price alerts is shown in Figure 8.5.

8.7 Field Polling and Data Gathering

One of the most innovative features of the Esoko Mobile Platform is its ability to facilitate the gathering of data from the field knowledge of what is happening to your suppliers, customers or beneficiaries can be both challenging and rewarding to businesses. Esoko uses the mobile platform to undertake SMS pooling with the creation of the required question or series of questions which are then scheduled to be sent to the targeted groups or individuals. For example, you could ask 'have you planted yet', or 'how many visits did you receive this week', or 'rate your satisfaction with the program 1–5'. The SMS responses received by Esoko are automatically arranged for reporting and analysis. The respondents and their answers can be viewed in real-time. The responses could then be grouped, sorted, plotted on maps or downloaded in various forms for deeper analysis. The company intends to improve this service offering with the introduction of value added features such as triggers and airtime incentives.

8.8 Bids & Offers

The EGCI provides the opportunity of any signed up user being an individual or a member of or a group to upload a bid to buy or offer to sell onto the Esoko mobile platform. This is done through simple SMS syntax such as 'buy 1mt maize'. The bids and offer to buy uploads can also be done through the web, or through a Java and Android mobile applications. A client can specify the offer price, origin, documentation, variety, source and many other relevant information of the merchandise. Responding offers are then posted into the client's network space immediately upon upload and are flagged as 'pending'. The client's network manager is responsible for reviewing the offers to remove a pending status. Using the automated alerts provided on the platform, a client could set up and receive automated SMS alerts whenever an offer is posted that matches specified criteria such as the commodity, location, size etc.

This is a simple matchmaking service that helps connect people. Through this platform, rural farmers are offered opportunity to aggregate their produce at the farm level and notify markets about the availability of their produce with the relevant specified information. Buyers are also offered the opportunity to advertise their demands and also access offers available if they match their demands and specifications. This service on the platform offers an important market connection and information which saves farmers a lot of money from post-harvest loses as potential buyers receive up to date offers. Farmers and sellers are also able to follow the price trends of their produce to ensure effective pricing and profitability. Buyers on the hand are given the opportunity to compare offers and prices and location to make informed choices as where to buy to ensure profitability. The platform thus create effective introductions as well as expanding both seller and buyer options.

8.9 Prices Mapping Services

The EGCI platform facilitates the provision of maps displaying prices, offers, and people as well as the polling results in a chosen location with detailed information on each location made available upon rolling over the map icons. All Esoko users can be identified through their location, enabling the tracking of how the members of a particular network interact with the system. For instance, the responses to a poll setup to identify users who have planted a particular crop could be analyzed through the generation of a map to identify who has responded and from where. A particular farmer group or individuals who have not planted the designated crop would be known and the required SMS alert sent to them. These maps are also available on the Esoko Android application, which will soon come with GPS capabilities so that user locations can be pinpointed, track the point where the user was profiled, and even plot the size of their farm, compound or factory.

Figure 8.6 shows a display of price mapping on a mobile phone.

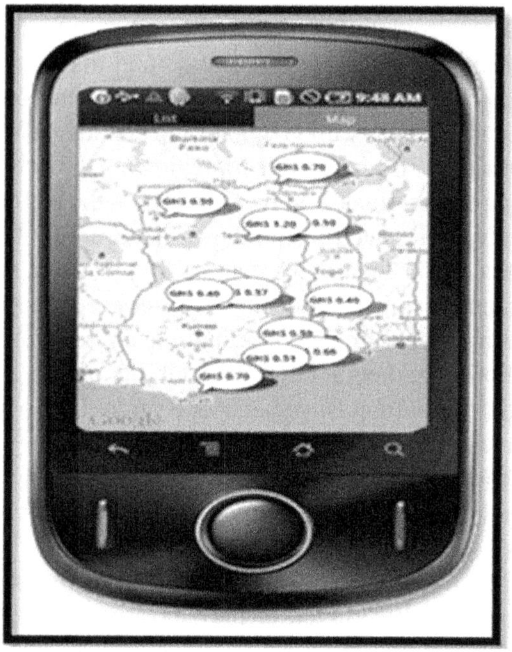

Figure 8.6 Esoko price maps interface

8.10 Impacts of the Esoko Agricultural Commodity Index

The Esoko Agricultural Commodity Index is providing an innovative marketing service that allows players in the agricultural value chain to access market information over mobile phones. With such extensive and up to date market information on mobile platform from across the country and beyond, producers, buyers as well as agricultural policy makers are armed with a potent and ubiquitous tool to access and compare prices, browse bids to buy or offers and product availability as well as their locations. Individual Farmers and farmers' groups can advertise their own products and services. This has resulted in improvements in revenue since players in the agricultural supply chain are able to negotiate better prices, or select more profitable markets for their produce (http://www.esokonigeria.com/newsreader.asp?news=2, 2013)

In Ghana, Esoko in partnership with the largest mobile network operator (MNO) in Ghana, MTN has expanded the platform to offer a lot more farmers the opportunity to sign on and user the platform in a program dubbed "Farmers First" (Bartlett, 2011). This partnership has resulted in the promotion ofThe EGCI platform in and around 15 well knows agricultural markets by proving training and sponsorship to interested rural famers to access price and extension information on their mobile phones whilst enumerators have been registered to collect data and upload on the platform (http://www.esoko.com/about/clients.php#casestudies, 2012). Twelve commodities are included in the index; groundnuts, white maize, gari, millet, soya beans, both imported and local rice, cowpeas, cassava tuber, wheat, tomatoes and yams. Markets were selected based on their regional location and strategic importance, and they include Accra, Bawku, Kumasi, Tamale, Techiman, Takoradi, and Hohoe (Bartlett, 2011).

The Esoko-MTN partnership has also registered 500 farmers from three districts to pilot the program (Bartlett, 2011). The EGCI platform delivers personalized price alerts, buy and sell offers, stock count and transport information on demand through SMS. Esoko has gathered information from 36 markets across Ghana and is located in 15 other African countries facilitating better control over prices of their products with mobile-mediated negotiation power resulting from market knowledge. There are over 10, 000 farmers utilizing the services in Ghana alone (Bartlett, 2011).

The reported success of Esoko mobile platform has also attracted the attention of reputable research institutions and universities, such as New York University, Georgia Tech, and International Organization's such as CIRAD and IFPRI (Subervie, 2011) and UNDP (David-West, 2010) who have and continue to conduct studies on the platform and the impact of the resulting increased transparency and access to market information on small producers and their families.

Private companies like Prestat are also supporting Esoko through corporate social responsibility initiatives to provide continual market and climatic information to farmers in the cocoa sector (Keeling, 2010), while MNOs like MTN are working with Esoko to bring a farmer club and agricultural voice helpline services to thousands of smallholders in Ghana (Boafo, 2009) (GNA, 2013).

Although the EGCI platform facilitates effective delivery information for farmers, the commercial nature of the platform is shifting the attention from important information such as climatic information to market driven product prices. The Government and NGOs interested in the impact of climate change on agriculture could therefore provide sponsorship to enable famers to access important climatic information to improve agricultural productivity in the country.

8.11 Challenges of the Esoko Agricultural Commodity Index

The process of identifying and training of enumerators for the collection of data on price and product availability has been a challenge for the Organization. This results in the data for price and product availability often being outdated thus generating complains from the partners in the agriculture supply chain on the unreliability of the information received.

Again, many of the partners and the resellers often underestimate the work involved in the promotion and reselling of the platform to potential users. This results in a number of partners not delivering the required services or dropping out altogether.

The market information is delivered in English to the farmers, majority of whom most are illiterates. Thus farmers are often asked to report for training with assistants who can read and understand English. Most of these assistants are their children. To eliminate this challenge the company has set up a call center where farmers can call and have messages, interpreted to them in local languages.

Challenges with the quality of serve provided by the Mobile Network Operators often results in the late delivery of information making the stalled information less useful. At the time of writing, the partnership agreement with a major Mobile Network Operator (MNO) in Ghana had been terminated and a new negotiation is on-going with another Mobile Network Operator (MNO) to provide the needed partnership.

8.12 Conclusion

The infusion of technology into agricultural extension in the areas of marketing and climate information dissemination is critical to improve the livelihoods of the millions of smallholder farmers across Africa. Improvement in general agricultural production, productivity and sustainability however depends on farmers' willingness and access to new technology. In many countries low agricultural production has been attributed, among other factors, to poor linkages between Research-Advisory Service-Farmers and to ineffective technology delivery systems, including poor information packaging, inadequate communication systems and poor methodologies. The unprecedented growth of mobile technology in Ghana with phenomenal mobile subscription exceeding 100% of the total population provides a platform for information dissemination to smallholder farmers.

A practical application mobile technology of tremendous use to smallholder farmers is the Esoko Ghana Commodity Index (EGCI); a mobile-enabled, cloud-based service which could be used on a basic mobile phone or a computer with SMS and data connection functionality. The EGCI has potential to transform agricultural extension. It offers its registered customers up to date marketing information. The EGCI platform enables the delivery of both wholesale (EGCI-W) and retail (EGCI-R) prices of agricultural products and their locations real-time. The players of the agricultural supply chain who receive this information are thus able to make important marketing decisions. It has also transformed the traditional mode of field data collection in Africa that involved physical visits to targeted area and asking questions, sending out questionnaire or making phone call. Through Esoko's innovative mobile platform, data is obtained through scouts who inform the managers at Esoko to setup automatic polls from chosen participants. Simple questions are then sent out via SMS at scheduled times.

Another critical area where Esoko's platform holds a lot of promise is in the arena of climate information dissemination. Through partnerships with National Meteorological Agencies, timely information could be made available to smallholder farmers via SMS who otherwise would have to access such informing via television or radio. When properly coordinated, the platform could play a vital role in the area of hydro-climatic disaster management as an early warning platform.

9

Harnessing ICT for Local Government Administration in Africa: A Look at the Push-ICT Theory Approach in Nigeria

Wilson Joseph,
Department of Mass communication,
University of Maiduguri, Nigeria
joeweee2003@gmail.com

Nuhu Diraso Gapsiso
Department of Mass communication,
University of Maiduguri, Nigeria
ndgapsiso@yahoo.com

Musa Usman
Department of Mass communication,
University of Maiduguri, Nigeria
almusa3@gmail.com

9.1 Introduction

The world is experiencing unprecedented changes as a result of the impact of Information and Communication Technology (ICT). In the last two decades, the world has witnessed an extensive adoption and use of Information and Communication Technology (ICT) for diverse purposes. Governance is one of the several areas that the impact of ICT cannot be overlooked. ICTs are already widely used by government agencies at various levels (national, states and local governments) to enhance the performance of government and bring it closer to fulfilling citizens' expectations.

With 774 Local Government Areas and about 160 million people, service delivery by the Nigerian government obviously would require several initiatives. This chapter explores how ICTs had been harnessed in Local Governments in Nigeria so far; how they were achieved and the possibilities the Push-ICT theory perspective offers in harnessing ICT in local government administration using Nigeria as a springboard to other African countries.

It is noticeable that effective administration greatly enhances performance irrespective of the size of an organization. Hence organizations (both government and private) are compelled to explore various avenues to ensure a meaningful and result-oriented administration. ICT has emerged in recent times to prove useful in facilitating the realization of administrative goals.

Guchteneire and Mlikota (2007 note that evidence from numerous projects and initiatives worldwide (eg. European Union: Connecting Europe, EU Cloud computing, United nations ICT4Development, International Institute for Communication and Development ICT projects in Developing Countries, International Development Research Centre ICT initiatives globally, etc) show that, effective utilization of ICT offers new possibilities for improved efficiency in governance, new ways of citizens' engagement and a more active participation in policy-making, resulting in improving and re-building of trust, transformation of relations between governments and their citizens.

There are people and communities that are often isolated and do not know their rights and options, lacking basic knowledge about the political and development processes that shape their lives. The absence of access to knowledge reinforces the vulnerability of many people to challenges, which forces them into exclusion, powerlessness, and other poverty traps. However, Singh, (2004) notes that ICTs are powerful tools when used in the right way as part of an overall development strategy to bridge the isolation. In the same vein Boating (2007) points out that ICTs are not the cure-all to the world's problems butthey can be vital tools to facilitate and enable affordable solutions for basic human needs and development. These technologies have made political, cultural and socio-economic integrations and networking much easier and faster than before.

Because ICTs have become an integral part of social, economic, and political interaction, countries worldwide are deploying ICTs for various developmental efforts. In this digital age, the role of technology in improving the lives of the people cannot be underestimated. Most people, more than ever are now buying goods and services online, sending messages across the globe to friends and relations, soliciting support via emails from donor agencies and receiving instant feedback (Ebeling, 2003).

However, effective utilization of ICT for effective administration and result-oriented governance cannot be realized by a mere wish. It involves rethinking, reorganizing and reprocesses that would influence behaviour and dispositions toward ICT for administrative purposes and governance so that government services at all tiers are delivered more efficiently to the people who need to benefit from them.

Efforts in many developing countries shows that ICT adoption and utilization is no longer a confined to the develop countries. In several developing countries ICT status is undergoing a rapid transformation especially in the area of computing and telecommunication. For example, Nigeria is no longer laggards as far as ICT are concerned. Way back in March 2001 the Federal Executive Council approved a National Information Technology Policy and the implementation of the policy started in April of the same year with the establishment of the National Information Technology Development Agency (NITDA), charged with the mission of making Nigeria an Information Technology (IT) capable country in Africa and a key player in the Information society, using IT as the platform for sustainable development and global competitiveness. (ICT4D Annual Report, 2007).

Furthermore, its liberalization drive has spurred initiatives and investment in the field of ICT. For example a Nigerian company Globacom has distinguished itself by registering its impact as a global telecommunication operator. Nigeria is giving back to the world (Daily Trust, 2009). From 2001 when the mobile telecommunication began a considerable penetration has been realized. Currently, Nigeria stands as one of the biggest and fastest growing telecom markets in Africa, attracting huge number of foreign investment. It is the continent's largest mobile market with now more than 110 million subscribers and it is estimated to rise to 120 million by the year 2014 (Internet World Statistics, 2011, Longe, 2013). The country has grown from one telecommunication (Nigeria Telecommunication: NITEL) to several telecommunication giants such as Globacom, Airtel, Etisalat, MTN, Starcomms, Multilink among others, and yet market penetration stands at only around 70% in early 2013 (Longe, 2013). It is also worthy of note that Nigeria internet usage status has improved. The population estimate of Africa in 2012 was 1,037,524,058, out of which there are 167,335,676 internet users. Nigeria has the highest number of users put at 48,366,179 (Internet World Statistics, 2012). Furthermore, the country leaped towards space science or satellite technology which it flagged off by the launch of two satellites. In 2003, a reconnaissance satellite: launched from the Plesetsk Cosmodrome in Russia and in 2007, the Nigerian Communication Satellite NIGCOMSAT-1: is expected to offer broadcasting, phone and broadband internet services for Africa (BBC World Service, 2007). These are all pointers to the fact that Nigeria certainly has a story or two to tell on the impact of ICT on its endeavors as a nation.

Governance as a key component nationhood, at whatever level is designed to better the lot of its populace and improving standard of living at each moment in time. Hence it requires constant reviews of administrative strategies to meet these mandates. Innovations are indispensable in this regard, to allow for administrative competence. Adoption of E-governance is one of the current initiatives in the global administrative drive, which is propelled by ICT. Thakur (2005) notes that E-governance is application of ICT for interaction among government, citizens and businesses, as well as in internal and external government operations to enhance governance. Adeyemo (2011) notes that E-government centers on interaction between Government and Citizen or Government and Customer; Government and Business; Government and Government; Government and Employee. The use of ICT for administrative purposes has been appropriated in developed countries like Britain, United States of America, France, Germany, etc. Thus Thakur (2005) points out that developing countries need to be pro-active and enterprising because of their presence in a digital loop. Putting in place IT policy by Nigeria is an indication that there is no receding on its effort to adopting Information and communication technology for governance and other developmental purposes.

The Local government, anywhere in the world is an important segment of governance structure designed to impact more closely on the citizens. It is widely described as the tier of government closest to the public and saddled with responsibilities that have direct impact on the public. However, it is observable that service delivery of Local governments in developing countries like Nigeria falls short of citizens expectations. Among other challenges, manual operation is still the "preferable" order of the moment, which obviously affects productivity and reduces output. High level of paper work is still visible in local government offices in Nigeria. "Service delivery to citizens is more towards office consumption less citizen centric, no significant re-engineering, no online workflows, etc" (Thukar, 2005). These often results to unnecessary delays and other complexities in the local government administrative system which affects its performance.

9.2 ICT for Administration

The role of ICT in any level government administration revolves within the principle or notion of e-governance. That is deploying and applying information technology for the purposes of facilitating daily administrative routine and duties of administrators. This also entails e-administration. E-administration is the conversion of paper processes to electronic processes. The major objective is for transparency and accountability, leading to better governance. Electronic services have the advantage of enabling the citizens to obtain information and to carry out transactions 24 hours a day, seven days a week, and are particularly suitable for simple administrative transactions, such as requests for permits, or submissions of tax files governance (Guchteneire & Mlikota, 2007).

Introducing ICTs in local government administration is equally introducing e-governance at the local government level. E-governance at this level of government is to improve management efficiency and service delivery in Local Governments, to increase Local Government Capacity and to create a unified e-governance system, where all administrative functions and obligation are integrated are harmonized for efficient service delivery (Samadashvili, 2011). Similarly Babatunde et al (2012) noted that:

> The ‘mainstreaming’ of ICTs within planning and design of development strategies helps to strengthen the establishment of efficient, effective and transparent governance systems. On-line tools can significantly improve the rendering of services and information flows from administrations to their constituencies. It also enhances communication at all levels as well as offering unique opportunities for broadened citizen involvement and participation in the decision-making process in standard and genuine manners.

ICT are not tools from out of space, they are all around us and are used for various purposes daily. ICT is the combination of Informatics technology with specifically communication technology. Information and Communications Technology (ICT) is a term that includes all technologies for the manipulation and communication of information. ICT encompasses any medium to record information (magnetic disk/tape, optical disks (CD/DVD), flash memory, etc.); technology for broadcasting information - radio, television; and technology for communicating through voice and sound or images - microphone, camera, loudspeaker, and telephone to cellular phones. It includes the wide variety of computing hardware (PCs, servers, mainframes, networked storage), the rapidly developing personal hardware comprising mobile phones, personal devices, MP3 players, iPad, etc. The Internet is another example which stands out as a major driver of most of the ICTs.

Administration entails the process or activities of running an organization. It is also the management of public affair. The local government is a tier of government that engages in activities that concerns the affairs of a public in a geopolitical entity. The importance of the Local Government is such that Awotokun (2005) points out that in Nigeria, the Local Government is not only accorded a place of pride in the socio-economic well-being the country, it is also seen as a way of bringing government closer to the people. Consequently, a uniform system known as single tier structure was adopted throughout the country. This uniformity can be conceptualized in terms of:

(a) The functions of local governments
(b) The structure of the local governments
(c) The financial resources of the local governments
(d) The place of traditional institutions in the local governments
(e) Relationships with state government
(f) Law enforcement.

The Local Government has constitutional responsibilities and has a mandate to perform several functions cutting across education health, infrastructural development, social welfare, etc. Hence, running a local government requires good administrative skills and in recent times adoption of information and communication technology. ICT made possible the emergence of service delivery models such as departments going on-line (Standalone/ different units interconnected, bottom up development takes place), conveniently located Service Centres (Counters manned by Government functionaries run by public/private agencies, multiple services at each location: payment, licenses, certificates, Xeroxing, downloading), self-service through a Portal single window interface (Paper forms and movement replaced with electronic forms, online interaction, transactions, integration for work flow and data sharing (Thakur, 2005).

Jensen (2002) notes that Local Governments are a fertile ground for the application ICTs because they are at the front lines of government in their service-oriented interaction with the public and business. As such, Local Government have considerable potential to assist in the process of integration of ICTs into the daily lives of managing its public affairs.

Thukar (2005) identified some areas where ICT come handy in administration:

- Support, simplify and connect government, citizen and business
- Enhance services delivery and empowerment of people by information available to people

- Availing latest information about global and domestic markets which facilitates transactions among businesses
- Availability of information and knowledge to policy makers which facilitates correct decisions related to grass root developments (bottom up, low cost solutions for effective planning and implementation)
- Overall enhancement of administrative efficiency and effectiveness of government processes

Jensen (2002) classified Local Government operations from an information technology perspective into three areas:

- Internal
- Intra-governmental
- External (with the public)

Many of these operations benefit considerably from the use of ICT. For example in Zanzibar Municipality in Tanzania, Lusaka municipality in Zambia, Maputo municipality in Mozambique, Abuja Municipal Area Council in Nigeria, etc use ICT facilities for these purposes.

ICT if properly harnessed would be useful in Local Government Administration in the following areas:

Internal Operations: internal functions that could use ICTs to assist in operations would include:

- Budget planning, (spreadsheet)
- Accounting and payroll operations
- Job costing and purchase/work order
- Geographical Information System mapping (planning managing and analysis)
- staff support (training, schedule management, contact management, email, web access)
- Monitoring and management of several Local Government affairs such as waste management, property management, Transportation (vehicle and fuel) management, etc.
- Electronic decision-support systems. (Local election and votingExternal Operations (with the public): ICTs can be employed to facilitate
- Easy interaction with the public
- Creating database system for various forms of documentation (plot'allotees', trade licenses: (monitoring expiry dates of licenses and tax to enhance timely revenue generation)

- Provision of application forms for various services over the web (land titles, subdivisions and zoning applications, building permit applications, property rent applications and payments, tenders

Underlying many of these uses of ICT in Local Administration, is the Internet which serves to provide the common protocol for many of these interrelated ICT functions to communicate, both internally and externally to other levels of government, business interests and the public. For example in some countries, such as South Africa, municipalities have a common payroll and accounting system administered centrally and processed locally (Jensen 2002). ICT can be adopted by Local government to produce a computerized land registration, creating an integrated system for the administration and management of land (eg. Abuja Geographical Information System), accessible through the Internet, encompassing all the geographic information that provides the process for authorization of requests related to the right to use and benefit from land. Through this system, it will be possible to provide information about the economic and legal situation of land, the types of permitted occupation, use and exploitation for various purposes such as agriculture, residence, industrialization and reserve (Jensen, 2002).

Appropriating the use of ICT by local government's administrations in developing countries would enhance the level of service delivery and staff performance to meet current global standard. Best practices in some countries have proven the efficacy of ICT in local government administration. For example in India, there are success stories in e-governance efforts: Registration of property deeds in Andhra Pradesh, Akshya Kendra in Kerala for delivery of services in rural villages/towns, updating of Land Records in Karnataka and Tamil Nadu, Community Information Centres for facilitating citizen services in 487 blocks on N-E States (Thakur, 2005). In Latin America and the Caribbean, the political engagement to promote the use of ICTs to enhance sustainable local development and enhance good governance is vehemently pursued. In the Latin American region a network has been created to connect municipalities throughout the region. The "i-Local" network launched by UNESCO, in collaboration with the University of Colombia allows sharing of knowledge and experiences among experts formed within the framework of specialization courses in local electronic governance. ICT in Local government participation can foster citizen consultation and participation in policy-making, by providing new possibilities for citizen involvement. "Today I decide" (TOM) portal launched by the Estonian government in 2001 is a significant instanceon the use of ICTs for civic consultation. This initiative provides an opportunity for Estonian to become involved in policy-making and to comment on draft laws that are published on the portal. The public can also submit their own proposals for laws or policies, which are taken into consideration by the government (Guchteneire & Mlikota, 2007).

9.3 ICT for Administration in Local Government in Nigeria so far/ Initiatives

Considering the importance of this tier of government the adoption and application of ICT in diverse areas of its services to its people has become imperative and calls for new approaches to ensuring effective incorporation of ICT in Local government administration. With a total of 744 Local Government Areas and a growing population of over 160 million it would not be fair to say there are no traces of ICT-propelled initiatives in Nigeria which has a direct effect on administrative performance at the grassroots. Reports and studies (e.g. Fantsuam Foundation, 2007, Ogbomo in 2009, Babatunde et al in 2012 etc). have shown that these technologies are used in Local Government Areas in Nigeria to improve administrative responsibilities.

A survey of 463 out of 774 Local Governments Areas across the six geo-political zones of Nigeria (South-East, South-South, South-West, North-Central, North-East, North-West) by Babatunde et al (2012), found that ICT facilities were used for various purposes and had different levels of ICT engagement. While some had high ICT engagement level (computerized staff records and payment system, high level mobile communication, availability of internet access, etc., some were very low (low level computer access, low level mobile communication for official duties, and no or grounded internet facilities, etc.). The ICT used in most of these Local Governments are: Computer, Mobile phone, Internet, Television, and Radio which are not different from Ogbomo (2009) ICT usage survey in Oshimili North Local Government Area of Delta State, Nigeria.

Babatunde et al (2012) looked at the performance level of the various ICT facilities. ICT for communication tops the performance level in most local government areas in Nigeria. The use of mobile phones and the internet for various degree of communication among major stakeholders: staff, management and officials of state government agencies, development, welfare and social services (using ICT for promoting information on people oriented services such as agricultural, health, legislative, loans and advances) commitment and record and data management among stakeholders. Financial services (use of ICT for funds disbursement, claims, grants and salary processing). social harmony (ICT is being used for promoting welfare and cordiality or harmony among stakeholders) Record and data management (ICT in areas of record keeping as well as management of different forms of data from various sources) Staff Development (use of ICT for seminars, conferences, staff training, recreation and various forms of entertainment) (Babatunde & Etal, 2012).

The efforts to achieve an ICT-driven society in Nigeria is not without challenges but there are also several initiatives designed to ensure affordable access for Nigerians. These ICT efforts at the Local Government and other levels in Nigeria are often made possible through various means which cuts across individual, Non-Governmental Organizations, private sectors such as financial institutions, telecommunication companies and the government.

ICT effort readily referred to is that of Association of Local Government of Nigeria which has, through its members (Who are Local Government Chairmen) from time to time procure ICT facilities in their Local Government Secretariats as the need may arise. States Ministries of Local Government Affairs in most of the states of the federation procure and install ICT facilities in the various Local Government Areas. For example in Adamawa State one ICT facilities available in all the Local Government Areas is Laptop computers donated by the state government through the Adamawa State Ministry of Local Government Affairs to key Local Government staff. Some few years ago, Internet facilities were installed in all the 774 Local Government Areas by the Association of Local Government of Nigeria (ALGON). However, most of them are no longer functional. A survey conducted through the Local Government Information Officers in Adamawa State (North East Nigeria) found that Michika LGA have functional Desktop computers in the Finance Department. Maiha LGA have Functional Desktops computers in the Admin and the Finance Departments, Madagali LGA have Desktop computers in all the Departments but only one functional is in the Finance Department. Ganye LGA have functional desktop computers in Admin and Health Department. The one in Health Department was donated by UNDP. Toungo LG. Admin and Finance Dept. also have computers which are functional. Numan LG also has a functional computer in the Finance Department. In addition to the above they also have digital camera both still and video bought by the LGAs. They also have cell phones and intercom.

Another ICT effort or initiative is the Universal Service Provision Fund (USPF). The USPF is a special fund set up by the Federal Government under the National Communications Act 2003. The fund is sourced from all licensed telecoms operators. They are required to contribute 2.5% of their annual financial turnover to the Fund, which are then used accelerate the deployment of ICT services to Nigerian, bridge digital divide and to compliment Nigerian Communication Commission and other stakeholders efforts. For example, NCC Wire Nigeria WiN and State Accelerated Broadband initiatives (aimed at linking the country's states with fiber optics cable and providing wireless broadband services in Nigeria). An ongoing USPF funded project by is Fiber Interconnection, in which MTN is taking fiber from their point of presence in the urban areas to the rural areas. Nigeria has acquired a fibre optics telecommunication infrastructure measuring 41,000 kilometers and about 11, 00 kilometers of the fiber optic cables network was laid between 2010 and 2013 out of which 4000 of this capacity is terrestrial cable along power transmission line. (Dada, 2007 NCC, 2012, Biztech Africa, 2013, Aamefule, 2013,.) Similarly the Galaxy Backbone Company owned by the Nigerian government is supporting deployment of VSATs (satellite terminals) across Nigeria. According to Dada (2007) the VSATs deployment would undoubtedly provide access to remote locations that are underserved and ensure that each of the 774 local Government Areas in Nigeria would enjoy connectivity.

Recently, Galaxy Backbone Company emerged the first in Africa at the 2013 United Nations Public Service Awards (under the category of "Promoting whole-of-Governance Approaches in the information Age") it was in recognition of its effort in the deployment of Information Technology in delivering services using the 1-GOV. net (a common platform built for all Federal Government Ministries, Department and Agencies to foster efficient and faster service delivery) (Goke, 2013). This is an indication that the initiative is still active.

A recent development in ICT initiative that is of direct impact to the Local Government Areas in Nigeria is the establishment of Rural Information Technology Centers Project (RITCs) across the Nation as part of Millennium Development Goals (MDGs) and the National Information Technology Development Agency's (NITDA) target objective of providing Internet access to the undeserved communities which is aimed at stimulating the growth of ICT in the country. This is one of the ways ICT facilities are made available in the Local Government areas (see http://www.nitda.gov.ng/index.php/projects/past-projects/143) for details of benefiting Local Government Areas across Nigeria. Although, the government has applauded this initiative, there are reports of neglect in some of the Centres. In September 2013 it was reported that one of the multi-million Naira RITCs in Jahun Local Government Area of Jigawa State (Northern Nigeria) has "suffered such neglect that its expensive equipment is gathering dust and becoming rusty. That the "The C-Band Internet facilities planted in the Centre only functioned on the day it was commissioned" (Sahara Reporter, 2013).

Another initiative that is of great benefit to administration in Local Government Areas in Nigeria (considering Babatunde et al, (2012) high performance rating of mobile phone usage in the Local Governments) is the Rural Telephony project. In 2001 the Federal Government of Nigeria initiated the National Rural Telephony Project (NRTP) to take telephony services to the rural areas. The project was to cover 218 local government areas in the first phase and provide over 636,256 Code Division Multiple Access (CDMA) lines in the 774 local government areas and the Federal Capital Territory (FCT) in the second phase and subsequently cover the entire Local Government areas in Nigeria. The project which is worth 200 million US Dollars (32 billion Naira), however, had some challenges of continuity in 2012. The official position by the Federal government through its Ministry of communications was that the project had been granted as a concession to the private sector for better implemented and management. However, Emejor (2012) put it thus:

> Fresh insights have been provided on why the $200million (N32billion) National Rural Telephony Project (NRTP) contract, conceived by former President Olusegun Obasanjo's administration 11 years ago, may have been abandoned……. Paucity of funds as well as nonchalant attitude of government and her supervising agencies played major role in frustrating the project (Emejor, 2012)

In 2012, in line with the Federal Government rural telephony initiative, one of the mobile network operators in the country, MTN Nigeria in conjunction with Huawei Technologies enhanced its rural telephony project with new 350 base stations thereby enhancing rural telephony infrastructure in about 850 villages across the country (Oketola, 2012).

Similarly in March 2013 Microfone Telecoms Nigeria Limited in partnership Globacom a telecommunication company launched the Rural Telephony initiative designed to ensure that not less than 80 percent of Nigerians living in the rural area enjoy the grassroots telephony initiative. The initiative would see farmers, artisans, traders doing business in the rural areas using the cost effective phones provided by Microfone to boost their endeavors and to enhance communication with government authorities (Subair, 2013).

Computers for All Nigerians Initiative (CANI) is another government ICT efforts designed to boost computer penetration level in Nigeria and to enhance Nigerians' access to computer hardware. Launched in July 2006 and coordinated by NITDA as a public-partnership initiative whereby members of the public especially civil servants will be able to purchase computers and pay back the loan at a low interest rate companies involved are Microsoft, Zinox and Omatek. In relation to this project is an initiative by a government agency (Petroleum Technology Development Fund (PTDF)) that embarked on building and equipping computer centers in secondary schools and tertiary education ion institutions across Nigeria. Some of the schools located in Local Government Areas afforded the dwellers an opportunity to access and use these ICT centers especially for internet related services. For example Federal College of Education (Technical) located in Potiskum Local Government area served as a reliable Internet Centre for the host community and the surrounding communities.

In an effort to strengthen ICT base of the nation the Nigerian government adopted the e-governance strategy to improve and ensure high level use of ICT for governance. The Federal Government through the National Information Technology Development Agency (NITDA) created the National e-Governance Strategies (NeGSt) with the mandate to facilitate, drive and implement the Nigerian e-Government Programme under a Public Private Partnership (PPP) model.

National e-Government Strategies, NeGSt, is a Special Purpose Vehicle (SPV) created in March 2004 by the Federal Government through the National Information Technology Development Agency (NITDA). The target for this program include t: the Presidency, Federal Government Ministries, Parastatals & Agencies, Armed Forces, the Nigerian Police, Paramilitary Establishment, Universities & Other Institutions, Judiciary, Legislature, State Governments Ministries & Agencies, Local Government Council, Citizens and Community Members, etc. according to Olufemi (2012) the E-governance effort of the Nigeria government is to improve the low level coordination and cooperation within the Nigerian public administrative structure. If properly implemented it would centralize databases, reduce cost of governance and achieve common standards among several government agencies. "Nigeria could benefit from such cost effectiveness in governance and improved services would enhance decision making in governance when common standards and timely data resources are shared" (Olufemi, 2012).

These initiatives have obviously improved the ICT-for-administration base of the nation at all levels of government. In relation to this development Danfulani (2013) points out that in the last few years all federal government Ministries, Departments and parastatals have floated web-sites providing information about their vision, mission, staff strength programs and activities.

> The State and Local Governments have since followed suit by linking their ministries, departments and parastatals to the web. This has given them a clear understanding and easy communication channels within and outside their States or Local governments. (Danfulani, 2013)

Initiative such as the National e-Government Strategies has helped in improving record keeping and updates at the Local Government levels. For example it had been very difficult to authenticate the workforce figure of Civil servant at federal, state and local government levels in Nigeria. Payment of salaries were manually paid by cashiers in Local government secretariats (table payment) and thus perpetrating inflation of personnel cost. The introduction of the egovernance strategies with financial platforms like the Integrated Payroll and Personnel Information System (IPPIS) by the three tiers of governments across the country had drastically reduced the financial excesses. According to Danfulani (2013) the Federal Government in salaries of said none existing workers (Ghost workers) and uncovered about 45,000 ghost workers on the payroll of 215 Ministries, Departments, and Agencies.

There were similar efforts at the state and local government levels. State's Boards of Internal Revenue have developed efficient data base that monitor's individuals and cooperate taxes and knowing when they are due for payment. This has cut down on the carelessness, fraud. Chief Executive officers now have most of the revenue accounts tied bank alerts showing payments and withdrawals from such accounts and can as well stop illegal transaction using their mobile phones or computers. For example Governor of Gombe State (North East Nigeria) has in place a syndicate accounting system that linked accounts of all states ministries and parastatals to single alarm system that tells the Governor of withdrawals and deposit of monies through mobile phone alert system. The system has obviously reduced fraud among public servants (Danfulani, 2013).

Another interesting stride by the Federal government to enhance ICT penetration at the Local government level is the mobile phone for farmers' initiative by the Federal Ministry of Agriculture and rural Development. The Ministry of Agriculture and Rural Development, and the Ministry of Communications Technology have concluded plans to distribute 10 million mobile phones to small-holder farmers in 2013. The phones would be designed to carry features such as information on market prices of farm produce, extension workers and how farmers can access agricultural funds, information on climatic conditions, planting periods, etc. This ICT initiative is aimed at subsidizing the cost of major agricultural inputs, such as fertilizer and seeds. This initiative would indeed reduce the difficulties encounter in accessing information especially at the Local Government level. Although the initiative has been criticized by Nigerians as unrealistic, Nkemachor & Nnadozie (2013) note that he mobile networks technology is a unique opportunity to give rural farm access to information that could transform their farming experiences. Instant updates on weather, crop prices, inputs availability, market and financial information can considerably improve their productivity and negotiating positions.

There is no doubt that introduction of the various ICT initiatives have shown some level of successes. However, there are gaps that need to be addressed to be able to explore the full potentials of ICT in Local government administration.

Generally, there are challenges often associated with incorporating ICT in governance. Some of these challenges are: inadequate or lack of training and capital, limited understanding of the potential of technology, and a lack of clear business strategies, higher costs of ICT introduction due to the scale of public organizations, paper documents required for administrative approvals, confidentiality of information (Hull and Milne 2001, Olabode & Akingbesote. 2007 Buhalis 1996). These are challenges that have spanned decades and can be considered as over flogged especially in an age where technological alternatives and user friendly technologies abound.

Mere wish cannot guarantee e-governance neither can over dwelling on surmountable challenges spur e-governance. There are steps that must be taken to achieve an information technology driven administrative operation in local government. Some of these steps include:

Policy framework: This is an important factor for diffusion of innovation in any society or organization and includes putting in place policy that would encompass the objectives of the innovation, phases of the innovation, training program, infrastructural deployment, etc. For example Nigeria has an Information Technology Policy designed to ensure appropriation of ICT for global competitiveness and to facilitate development. Thus Local governments can further adopt the National IT policy and further design specific in-house policy to spur appropriation of ICT in its daily administrative endeavors. This important step iscapturedby Thukar (2005) (Thakur, 2005).

This position includes:

- Personal initiative of reform minded and technology savvy Civil servants,
- Continuous technology support of National or state information technology Agency (e.g. NITDA in Nigeria)
- Well defined objectives, guidelines and funding

a) Make available to citizens, through the Internet, the most sought-for information, including application forms and similar documents needed to be filled-in by the public.
b) Create a Competitive environment amongst departments: creating a competitive environment among departments and staff would encourage usage of ICT facilities. For example bonus of responsibility allowance can be increased all staff of a department that fully use ICT for its operations.
c) Provision of relevant infrastructure (automated offices and IT centres for members of the public), and create avenues for human resources development.
d) Encourage on-line payment through payment gateways and portal as a single-window for services from a large number of departments.
e) Subsidize or offer discounts for online dealings with members of the public. For example offering discount for members of the public that process applications online for land allocation, licenses, shop allocation, tenders, etc.
e) Up-to date information about services of the local government should be published on the web for download by members of the public. For example, applying for various certificates/licenses, other services such as scheduling appointments with the Chief Executive of the Local government and other services.

9.4 The Push-Theory Approach

The Push-ICT Theory according to Wilson (2009) is a derivative of the word "push". Which means to press until it's accepted. The theory is a product of several years of observation of the situation surrounding the use of some information and communication technologies (such as mobile phone, computer and computer literacy training).

The history of computer deployment in Nigeria dates back to 1960 but it did not gain popularity until the 1990s and the beginning of the new millennium (2000) when several organization and government agencies, educational institutions began a massive deployment of computers for various purposes.

Although there was enthusiasm among Nigerians but there was also resistance to change some old practices like manual record keeping, manual payroll for staff, manual processing of students' results and lecture notes, manual banking services, etc. However, the earliest push came from the Nigeria government through introduction the Information Technology policy (Policy Push) which it tied to every aspect of national development (agriculture, education, economy, governance) indirectly coercing Nigerians to adopt. Most organizations made computer literacy one of the requirements for employment, promotion, etc. The second push was government and organizations deployment of ICT facilities (Deployment Push). It deployed computers to government agencies and educational institution through organizations like the Petroleum Technology Development Fund-Nigeria, Educational Trust Fund -Nigeria. It also introduced the One -US- Dollar-Computer initiative. Computers were also made affordable to Nigerians through soft loans (payable in several months) these and other efforts pushed computers to the doorsteps of Nigeria and served as a kind of indirect coercion for Nigerians into the use of computers.

There was also social coercion or push (Social Push). Organizations (educational institutions and their proprietors, businesses and individuals made computer services and ownership, mobile phone ownership a social status symbol. This forced individuals and organization into buying computers, offering computer services and mobile phone to belong to the supposedly social class of these technologies users. The Social push could come from close associates (friends, colleagues. spouses, and people not known (competing organizations, similar businesses, etc.) to the user.

Another angle of the theory is the push from ICT users (Users Push). When ICT facilities are not affordable which in most cases is as a result of poor implementation of policy, the users of these ICTs are forced to complain to relevant authorities. In Nigeria the high tariff hitherto on mobile telecommunication services has compelled subscribers to push for reduction on tariff through the legislative arm of government and the Nigeria Communication Commission.

The Push-ICT Theory stipulates that:

- In a situation where Information and Communication technologies are considered important or relevant to development of individuals or community, such technologies should be deployed by the relevant organization (government, non-governmental organizations or individual.
- The relevant technology or services (e.g. training) should be made affordable. It can be deployed free or highly subsidized.

- The deploying organization or individual would clearly identifying workable benefits of the technology and subsequently coerce the individuals or communities to use the deployed technologies or services.
- The push is usually through policy framework, cheap and affordable deployment of ICT facilities, social status push and ICT-user push.
- Where ICT remains unaffordable (may be as a result of poor implementation of policy) users also push ICT providers to offer affordable facilities and services.

When there is easy access or availability of ICT facilities, resistance to adopt the use of ICT is highly reduced, while acceptance to use is greatly enhanced and the possibility to use the technology is high.

9.5 Push-ICT Theory and the Local Government

UNESCO project mission report (2002) points out that Local Government Areas in Africa are a fertile ground for the application of ICTs. They are at the front lines of government in their service-oriented interaction with the public and business. However, there are challenges hampering the full adoption of ICT for local government administration in Africa. According to (Ogbomo, 2009) some of these challenges, especially in the local government areas include power supply, illiteracy/ICTs illiteracy, low level of ICT skills, the high cost of ICTs, and lack of facilities such as cybercafés. These challenges would be greatly addressed if policies are well implemented. It is observable that the policies only exist on paper and little effort is made by the policy makers and local government administrators to provide an enabling environment for the use of ICT to enhance governance.

The Push-ICT Theory approach would enhance the implementation of the use of ICT in local government administration in African countries. First, Policy push is important in this regard. For example the reluctance by local government staff can be addressed by policy push. Training can be made affordable and mandatory for all staff. Staff can be sponsored to acquire computer skills. For example in The College of Education system in Nigeria, Academic staff are mandated to acquire computer skill and certificate or would not be promoted or not go beyond certain rank. A period of five years was given as deadline. However, Institutions in ensuring compliance, made affordable computer training opportunities available for staff. For example in federal College of Education (Technical) Potiskum (Northern Nigeria) There was a large number of application from within the institution (especially academic) for its 3 months Certificate course in Computer Appreciation The Large application made the Management restricted admission for the first intake to just staff of the institution and subsequently opened it up to all interested candidates. It is important to note at this point the ICT- deployment Push is necessary to achieve some level of ICT adoption. If necessary facilities (computers, internet connectivity, etc.) and services (training) are made available coercion to use the facilities would not be too much of a task to achieve. Coercion is impossible with the absence of ICT facilities and services. If facilities and services are made available at the Local Government level it would be easy to ensure that staff and members of the communities use the facilities. For example applications for land allocation, shops and other services of the local government can be applied online. In this case the members of the community

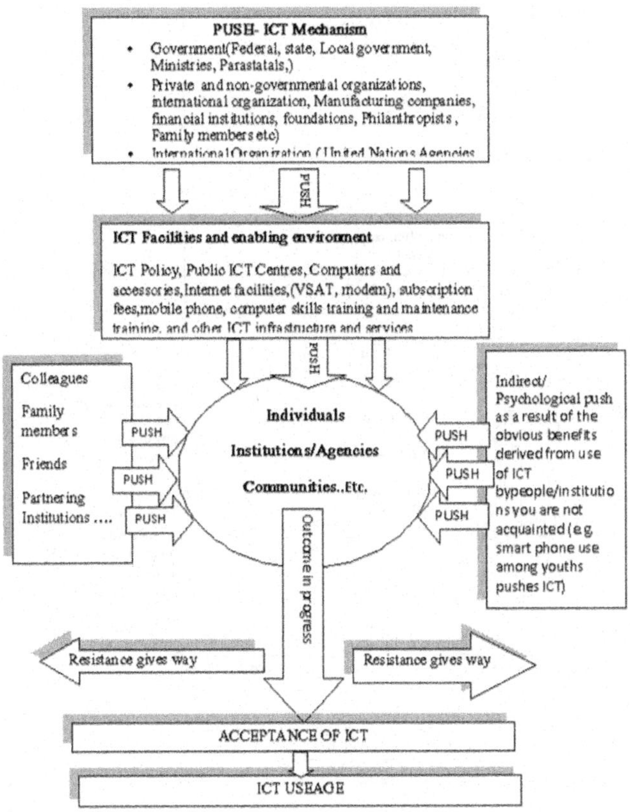

Figure 9.1 PUSH-ICT model

Source: Wilson, 2012

would be forced to use the ICT centre in the community. The accounting staff of the local government can be mandated to adopt electronic payment methods to avoid the current practice of moving huge amount of physical cash. Contractors and staff salaries can be paid electronically. This are attainable when facilities are deployed and users coerced to use them.

Push can also come from the users and communities (ICT-User push). Users can also push the local government administration to make alternative power supply available such as solar power and generators to constantly power the ICT facilities. Communities can press for ICT centers that would provide affordable services.

The figures below give a pictorial clarification of the Push-ICT approach:

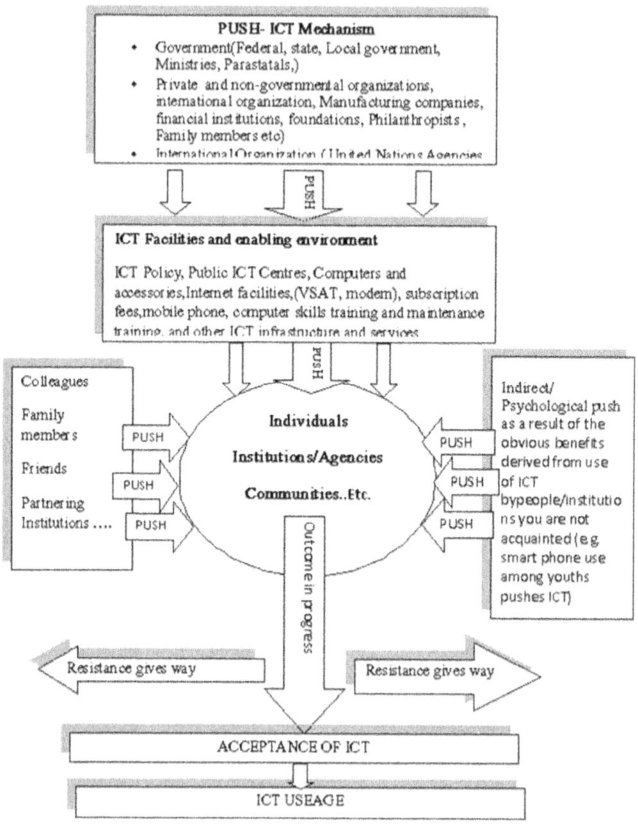

Figure 9.2 PUSH - ICT model applied to local government scenario

Source: Wilson, 2012

9.6 Conclusion

Every level of government strives to meet the yearning of its people. Thus it is imperative to explore avenues to achieve effective administration. In this digital age where the relevance and impact of any geopolitical entity is to an extent determined by its ICT capabilities, it is imperative for local governments especially in developing countries to appropriate ICT in its management system. Appropriation of ICT in local government administration would not only raise performance level, but would encourage citizen's participation in governance. The Push-ICT Theory approach would help make these technologies available and indirectly force members of the community to use them. The various ICT initiatives of the Government of Nigeria and others are obvious display of the Push-ICT Theory approach through the use of Policies and other forms of coercion to ensure that ICT are readily available and fully harnessed for administrative purposes or governance. Focus should be further directed to the Local Government Areas to ensure that Federal Government ICT initiatives are fully implemented and not treated with levity as the case has always been with government projects that are not backed up by intensive monitoring and evaluation.

10

Using Mobile Phones for Environmental Protection in Africa: The Equatorial Africa Deposition Network Case Study

Alejandro Islas Lopez
McMaster University
Canada
alex.islas@gmail.com

Gail Krantzberg
McMaster University
Canada
krantz@mcmaster.ca

10.1 Introduction

Because of their ubiquity potential, mobiles phones represent the most used form of telecommunication worldwide (Bhavnani et al. 2008). In the case of Africa, by providing communications channels that were previously nonexistent, mobile phones have revolutionized all kind of sectors (Ekine 2010). Under the context of a telemetry project aimed at investigating the atmospheric deposition of nutrients into the African Great Lakes, this chapter explores how mobile phones can be used for environmental protection.

In the first section, a short description of the African Great Lakes and their environmental and economic importance is presented. The second section provides an overview of the Equatorial Africa Deposition Network (EADN), an ambitious transboundary project that tries to identify sources of eutrophication[1], a major threat to the African Great Lakes integrity. In the third section,

[1]Eutrophication is defined as "the over enrichment of receiving waters with mineral nutrients. The results are excessive production of autotrophs, especially algae and cyanobacteria". (Correll 1998)

a mobile environmental framework around the EADN project is presented. The key role mobile phones can play as development boosters is presented. The mobile framework proposes services that can optimize the EADN daily operations and ultimately, presents mobile based solutions that can help to understand the behavior the communities that directly interact with the African Great Lakes have. Finally, the fourth section present recommendations towards the implementation and use of the mobile framework and general conclusions.

10.2 The African Great Lakes

Delimited in the north by the Lake Turkana Basin and in the south by the Lake Malawi Basin, the East Africa Rift Valley Region (EARVR) is a trans-boundary natural and human habitat that involves the following countries: Ethiopia, Kenya, Sudan, Uganda, Tanzania, Rwanda, Burundi, Democratic Republic of Congo, Zambia, Malawi and Mozambique. As shown in Figure 10.1, the EARVR includes lakes Victoria, Tanganyika, Malawi, Turkana, Albert, Edward, George and Kivu. These tropical lakes form the African Great Lakes (AGL) and act as sources of the Nile, Congo and Zambezi African rivers. Because they have surface areas and volumes comparable to those of the Laurentian Great Lakes of North America, it is common to find references in the literature of Lakes Victoria, Tanganyika and Malawi as the only members of the AGL.

10.2.1 Lake Victoria

This is the second largest lake in the world and the largest in the African Continent, Lake Victoria is bordered by Tanzania (51%), Uganda (43%), and Kenya (6%) (Odada 2006). However, Burundi and Rwanda are members of the lake Basin. Located at 1134 meters above sea level, the lake has a catchment area of 184000 square kilometers. Rainfall, evaporation and the outflow that comes from the Nile River are the major components of the Lake's water balance (Spigel and Coulter 1996).

With a population density that goes up to 1200 persons per square kilometer (Hoekstra and Corbett 1995), Lake Victoria basin support one of the densest, multi-ethnic and rural populations in the world. 80% of the lake catch is related to some kind of agricultural practice (Majaliwa etal.2000). As a result, 21 million of people in the Basin have agriculture as their main source of money, with an average income of 90–270 USD per year (World Bank 1996).

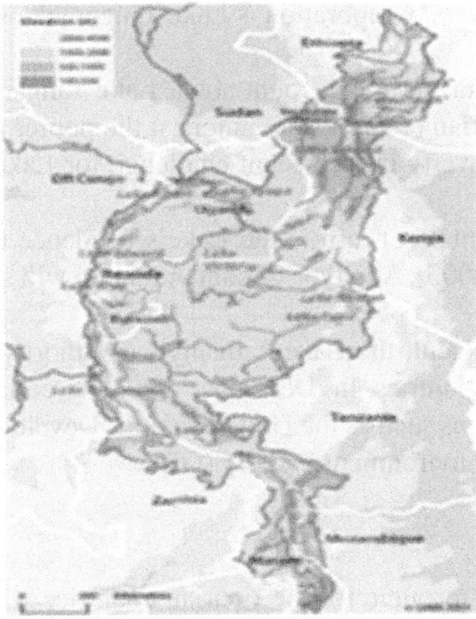

Figure 10.1 EARVR region (Odada 2006)

Besides agriculture, fishery and manufacturing represent the two other major economic sectors related to Lake Victoria. As expected, fish is the most affordable and common source of animal protein within the riparian countries (Bwathondi et al. 2001). It is estimated that the exports made by Lake's Victoria fishery industry represent earning of about 600 million USD (Duda 2002).

Although Lake Victoria is the major source of water and animal protein for the surrounding communities, the most common health issues are related to waterborne diseases. This can be explained by the fact that the lake is a repository for human, agricultural and industrial waste and taking water for domestic consumption directly from the lake without previous treatment is common practice in the lakeshore communities (Bwathondi et al. 2001).

10.2.2 Lake Tanganyika

With a length of 673 kms, Lake Tanganyika is the longest lake in the world (Odada 2006). Its shoreline perimeter is shared between Burundi (9%), Democratic Republic of Congo (43%), Tanzania (36%) and Zambia (12%). (Hanek etal.1993). Fed by a catchment area of 220,000 square kilometers, the Rusizi and Malagarasi are the major water inputs into Lake Tanganyika, along with numerous small rivers. Evaporation is Lake Tanganyika's main cause of water loss (Coulter 1991).

A fact that must be taken into account about Lake Tanganyika is that all of the surrounding riparian countries are amongst the poorer in the world. It is clear that extreme poverty is a constant challenge for Lake Tanganyika's riparian countries.

A major environmental concern is the Lake's residence time capability (400 years) (Bootsma, 2003). Any type of contamination will stay in the Lake for a long time.

In terms of management, the Lake Tanganyika Authority was launched by the four riparian countries in December 2008 in collaboration with international organizations such as the United Nations Development Program (UNDP) or the Global Environment Facility (GEF).

10.2.3 Lake Malawi

Lake Malawi is a great resource for the riparian countries when taking into account the drought-prone and semi-arid characteristics of Southern Africa. Even though it has an area of 126,500 km2, it only has a catchment area of 97,750 km^2, something not that common considering its magnitude (Drayton 1984).

With a fish fauna comprising 800 species, some of the endemic like the cichlids, Lake Malawi is the lake with more species of fishes in the world and, as a consequence, the most vulnerable to fishing pressure (Ribbink 2001). Such amount of biodiversity has led to international efforts to protect the lake's fauna such as the Lake Malawi Biodiversity Conservation Project.

According the World Bank (2003) 80% of the lakeshore population lives in Malawi. In addition, 70% of Malawi's land area is constituted by the Lake Malawi Basin. These statistics shows that any kind of human impact assessment on Lake Malawi must prioritize Malawi as the initial country subject of study.

Water-borne diseases like schistosomiasis are common in coastal settlements around Lake Malawi as a result of the lack of secure water sources. Even though all the riparian countries have passed policies related to issues around the Lake, they have been made without consultation and independently. Launched in 2003 and funded by the World Bank, the Lake Malawi Ecosystem Management Plan was the first multinational effort intended to maximize the benefits of the riparian communities while sustaining the Lake ecosystem (World Bank 2002).

10.3 Equatorial Africa Deposition Network (EADN)

10.3.1 Biomass Burning In Agriculture and Energy Production

As has been stated, the agriculture industry developed around the AGLs basins and catchment areas represent a significant percentage of the riparian countries' Gross Development Product (GDP) and labor force occupation as shown in Table 10.1.

Table 10.1 AGL riparian countries agriculture GDP and labor force occupation percentages

Country	GDP	Labour force Occupation
Tanzania	27.7%	80%
Uganda	24.2%	82%
Kenya	24.2%	75%
Burundi	30%	93.6%
Rwanda	33.3%	90%
Democratic Republic of Congo	44.2%	Not available
Zambia	85%	85%
Malawi	29%	90%
Mozambique	29.5%	81%

Source: https://www.cia.gov/library/publications/the-world-factbook/

Electricity access represent a constant challenge African population faces. By 2009, 585 millions of people in Sub-Saharan Africa have no electricity access (IEA 2012). In terms or electrification 30.5%, 59.9% and 14.2% represent the total, urban and rural rates in Sub-Saharan Africa, respectively (IEA 2012). Table 10.2 provides the electrification rate and amount of population without electricity access of the AGLs riparian countries.

Table 10.2 AGLs riparian countries electricity access.

Country	Electrification	Population without electricity in millions
Tanzania	13.9%	37.7
Uganda	9%	28.1
Kenya	16.1%	33.4
Burundi	Not available	Not available
Rwanda	Not available	Not available
Democratic Republic of Congo	11.1%	58.7
Zambia	18.8%	10.5
Malawi	9.0%	12.7
Mozambique	11.7%	20.2

Source: (IEA 2012)

In hands of farmers or housewives, biomass burning is a common practice in developing countries. Andreae (1991) recognize several purposes biomass burning serves:

- Land clearing for agricultural use, either for shifting agriculture or for permanent removal of forests. Usually, at least two burns of dry out vegetation are associated with land clearing (Fearnside 1985).
- Grazing and crops lands nutrients regeneration.
- Charcoal production for industrial and domestic use.
- Energy production for domestic use such as cooking or heating.
- Weeds and bush control in Savannahs, common ecosystems in Africa. In this case, biomass burning is executed to avoid the overgrown of grassy vegetation, a needed condition for grazing, and the most common agricultural use of the grassy savannahs. (Andreae 1991).

Even though the majority of emission to the atmosphere from biomass burning is dominated by oxides of carbon, relatively low levels of nutrient emissions will be present because of the presence of such elements in dry plant biomass (Bowen 1979). In terms of the amount and type of biomass burned in Africa annually Delmas and others (1991) found that 2.9 Gig tons of dry matter are burned every year and that savanna fires are the most common type of human related biomass burning emissions.

Studies like the one mentioned above have spotted savannas' biomass burns as one of the major sources of nutrient emissions in Africa. However, little information related to nutrients atmospheric deposition have been gathered. Funded by the GEF, International Water Projects on the AGL have enabled empirical measurements at lakeside locations that have thrown estimates about wet[2] and dry[3]3 deposition of nutrients. Due to the fact that management efforts such as the Phase 2 of the Lake Victoria Management Plan will propose more extensive monitoring of nutrients deposition, better understanding of transport macronutrients around the African continent is needed (UNEP 2011).

---

[2]Wet deposition is defined as "the is the transfer of atmospheric compounds to the earth's surface via precipitation" (Airzone One and Bootsma 2011)

[3]Dry deposition is defined as "the transfer of compounds to the earth's surface by uptake of gaseous compounds or the deposition of aerosols or particles" (Airzone One and Bootsma 2011)

10.4 EADN General Description

Research done by the International Water Project on the AGL has provided evidence of the relationship between atmospheric deposition of nutrients and eutrophication. The majority of these projects have been deployed around the lakes' shorelines or catchment areas (UNEP 2011). Projects aimed at investigating how much nutrients are transported from outside the basin areas into the lakes are needed in order to identify the largest sources of nutrients over enrichment that are already causing environmental damages to the AGL, being Lake Victoria the most affected.

In the context of a workshop organized by the African Collaborative Center for Earth System Science (ACCESS) held in Kenya in 2005, the need to design and implement a long range atmospheric deposition network that includes measurements out of the lakes' shorelines and basins areas was identified. As a result, the Equatorial Africa Deposition Network (EADN) project was proposed with the support of the following 12 countries: Burundi, Cote d'Ivoire, Democratic Republic of Congo, Ghana, Kenya, Malawi, Mozambique, Nigeria, Rwanda, Senegal, Tanzania and Uganda. The EADN project was designed in a way that will extent from areas where biomass burning is intensively practice in central and southern Africa, to the west of the continent in which Sahelian dust is expected to be a major source of Phosphorus. The declared objectives of the EADN are to identify the sources of atmospheric nutrients in Africa, trigger the mechanisms that introduce such nutrients to the atmosphere, determine the pathways by which nutrients are transport and finally, understand how much atmospheric deposition contributes to the AGLs eutrophication (UNEP 2011). Consisting of 12 points, Figure 10.2 shows the general locations of EADN's monitoring stations. Even though a detailed technical description of the network is out of the interest of this paper, its main characteristicsare presented as described in the EADN operational manual (Airzone One and Bootsma 2011):

- Two types of sites exist, regional representatives that provide estimates for deposition of nutrients (Nitrogen and Phosphorus) in regions inside and outside the AGLs catchment areas. Lake-side sites provide direct estimates of nutrients deposition within the AGL[4].

[4]The EADN operational manual provides the exact coordinates of 11 of the 12 monitoring stations. By March 2013, with 4 out of 12 monitoring sites installed, the network was not 100% operational. It is unknown by the author of this paper when will the totality of the sites will be installed

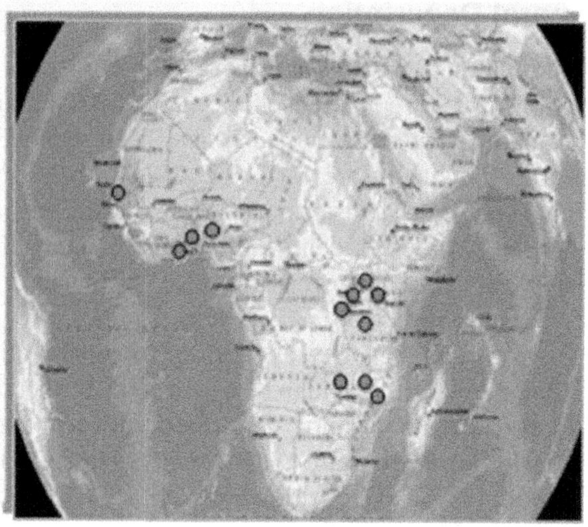

Figure 10.2 EADN's monitoring stations (Airzone one and bootsma 2011)

- To determine the airborne nitrogen and phosphorus compounds, the network includes active and passive measurements of airborne particles and of reactive gasses.
- The network has the capability to collect event-based precipitation.
- In order to support results modelling and interpretation, meteorological measurements will be provided.

10.5 Institutional Framework

Due to the fact that the mobile framework that the next section proposes was conceived as a complement to EADN's monitoring sites, it is important to identify the different stakeholders involved in the project, as they could be the one responsible of the operation of the mobile framework. Figure 10.3 shows EADN organization structure.

10.6 Mobile for Development

The Information and Communications Technologies for Development (ICT4D) movement is based on three major domains: Computer Science, Information Systems and Development Studies. As explained by Heeks (2008), this multidisciplinary approach tries to propose solutions that, while

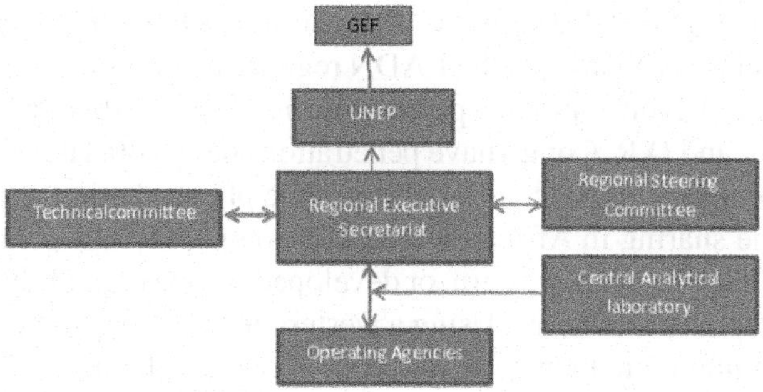

Figure 10.3 EADN organization structure (UNEP 2011)

still being techno-centric, incorporate other concepts like understanding the human, political and economic contexts around a specific development problem.

When thinking on the role mobile phones can have as part of the ICT4D movement, we have to consider the challenges such technology faces in developing countries. In terms of infrastructure, lack of stable power sources and mobile coverage are common issues, especially in rural areas because of the urban-first approach followed when deploying mobile services (Wicander 2010). With around 800 million people worldwide that are not capable to read and write, illiteracy represent a challenge when using mobile phones and also when trying to use mobile services like text messaging (Knoche 2012).

Figure 10.4 shows the growth trend mobile phones penetration have had in EADN countries in the last years. Actually, mobile penetration in Africa has

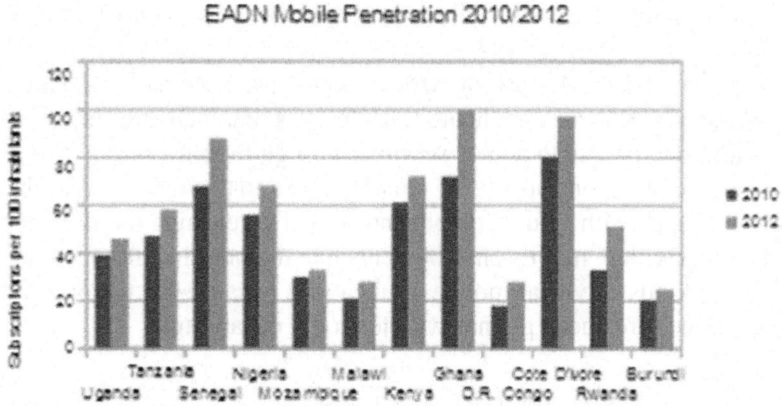

Figure 10.4 EADN penetration. (ITU, 2013)

doubled since 2008 (ITU, 2013). Figure 10.4 also shows the different realities the mobile industry has all around the EADN region, in countries like Ghana, Senegal and Cote d'Ivoire, mobile penetration is close to 100. In contrast Malawi, Burundi, and D.R. Congo have penetrations below 40. However, this statistics hide a common practice in Africa: mobile phone sharing.

Mobile phone sharing in Africa represents a completely different understanding of a mobile phone. In western, or developed countries, mobile phones are seeing as personal accessories. Using a western mentality, a mobile phone represents something more than a way to keep in touch with others. A mobile phone represents an accessory that clearly reflects the owner personality. Things like size, color or external design are always considered when buying a mobile phone. In rural Africa a mobile phone is not always seen as a personal accessory. Simply stated, a mobile phone is a way, and many times the only one, to connect outside the villages. Under this understanding, mobile phones are conceived more a community asset.

Because acquiring a SIM card is cheaper than acquiring a mobile phone, it is common practice that individuals have more than one SIM card, even from different cellular carriers, and make use of them by means of the community mobile phone. In low income/rural areas, it is expected that communities will have a limited number of mobile phones, usually one or two, which people share (Zuckerman 2009). For example, by 2006, a survey between households in rural Botswana revealed that 62.1%, 43.8% and 20% of the phone owners share their phones with their families, friends and neighbors, respectively. From these mobile phone owners, only 2.2% of them charge for the use other people make of their mobile phones (Gillwald 2005), fact that reinforces the idea that a mobile phone is more a community utility rather than a personal object. Regional illiteracy is something that must be taken into account.

According to UNESCO, 38% of African adults are illiterate[1]. Research like the one done by Knoche andHuang (2012) states that there are strategies that allows illiterate people to access services through mobile phones. Some of those strategies rely on the use of smart phones that would enable the development of apps with visual components aimed at reaching the illiterate population. As expected, mobile environments like the ones in Africa are not suited for these kinds of visual mobile applications because the majority of the existing phones are cheap phones with text only capabilities.

[1]http://www.unesco.org/new/en/dakar/education/literacy/

Two factors might help to break the illiteracy barrier when using mobile phones as part of the EADN. First, as stated above mobile phones are more a social than a personal asset in Africa. When using them, social interaction and cooperation are expected. Such community effort might be a crucial factor to help illiterate people access mobile phones services. Second, national and international efforts have reduced the illiteracy rates in younger generations. Adults that are illiterate can rely on younger members of their families when accessing text based mobile applications.

Besides relying on community efforts, Interactive Voice Response (IVR) systems can represent another way to efficiently deal with illiteracy. Those kind of spoken dialog systems have been successfully used in rural settings to allow communities with high illiteracy rates to access meaningful information to their everyday life like agricultural information (Gumede and Plauché, 2009).

Talking about the mobile industry environment, all of the EADN countries have more than one cellular carrier being MTN a case worth of mentioning. As Africa's leading Telecommunications provider, MTN is the major carrier in 5 of the 12 EADN countries and it's also the carrier with the biggest number of subscribers in one single African country, Nigeria (Blycroft 2012). Actually, the existence of multiple cellular carriers has been a crucial factor in Africa's mobile penetration rates improvements. In terms of pricing, one interesting thing to highlight is the competitive pricing ratio between one minute of voice and one text message (SMS). For example, by 2006 Kenya, Uganda and Tanzania had a ratio around 1:6 (Mendes et al. 2007). Without a doubt, this pricing condition has been a key contribution for the development of the current SMS culture in Africa.

In terms of continental cooperation, the 12 EADN countries are member states of the Africa Telecommunications Union (ATU), the umbrella organization in charge of promoting development of info-communications in Africa.

10.6.1 mEF Definition

As stated during the EADN description, identifying the sources of atmospheric nutrients in Africa is one of the project main objectives. Even though past research projects have already tried to identify sources of nutrients emissions at the lakes' shorelines and catchment areas, one unique characteristic of the EADN is that it will try to identify long range sources that could come into the AGLs region from distant sources like, for example, the Sahara or the Sahel. Once the nutrient emissions sources are identified, hypothesis like the one proposing biomass burning as responsible of the AGL eutrophication could be evaluated. Related to its social and economic environment, Africa is the continent where human related biomass burning is the greatest (Delmas et al. 1991). Because of that, any project intended to fully characterize nutrients emissions associated to biomass burning has to incorporate strategies aimed at understanding the human behavior around such activity. Recognized as the ICT with the biggest penetration worldwide and with an impressive growth in Africa for the last 10 years, mobile phones represent a great opportunity to interact with communities all over the EADN region. Likewise, mobile phones can be used for something more than human oriented purposes. With data transfer capabilities, mobile phones can also be used as the "last mile" communication channel in a data network. Such characteristic must be taken into account when no other ICT is available, a reality that is common in developing countries like those of the African continent. One last characteristic worth of mentioning about mobile phones is, precisely, its mobility and how it can be related to their users mobility patterns, an approach that can be useful for understanding any kind of human impact.

As explained before, a major challenge faced by the mobile industry of EADN countries is the lack of proper electricity sources. However, many of the Mobile Network Operators faced this situation by using other sources of electricity, like solar, to energize their mobile infrastructure. Although requiring more investments from the MNO's, unstable or nonexistent power grid infrastructures do not represent a barrier to access mobile services. For example, according to a survey made by Vodafone in 2005, 97% of the of people in rural areas that use biomass burning as their source of energy had no problems to use mobile phones in their everyday life (Wicander 2010). In terms of environmental protection, as suggested by Mungai (2005), mobile phones represent an effective channel for creating awareness, promote sustainable practices, strengthen warning systems and enable communication channels between environmental agencies. The African Great Lakes represent a complex management case when taking into account the amount of countries involved. Agencies like the Lake Victoria Region Local Authorities Cooperation, the Lake Tanganyika Authority or the Lake Malawi Evaluation Group must see on mobile phones a regional channel that can be used to spread a common message. Parallel to the telemetry network used by the EADN, a mobile framework could be implemented

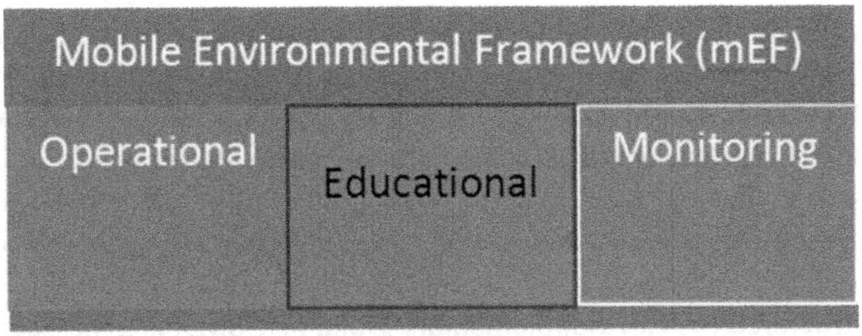

Figure 10.5 Mobile environmental framework layers

to analyze the human component of biomass burning and also to facilitate a more efficient exchange of data between the operating agencies and the central analytical laboratory. A mobile environmental framework (mEF) with operational, educational and monitoring components is presented in Figure 10.5. A more detailed description of each one of the layers of the mEF is discussed.

10.6.2 Operational Layer

As mentioned during the project description, the countries' operating agencies will the ones responsible of the day by day activities. More specifically, the EADN field operations manual describes the protocol Sites Operators must follow. The procedure includes precipitation and air samples that must be collected daily and weekly, respectively. As part of the package that must be shipped to the Central Laboratory with the samples is the Sample History Form (SHF), document that must be filled by the Site Operator and contains information such as meteorological or field chemistry data related to the time in which the sample was collected. Once the Central Laboratory receives the samples, the staff must enter the SHF data into the database to develop further analysis. Every month, the Site Supervisor and Site Operator will receive a report from the Central Laboratory that details detected problems or recommendations for improving accuracy. Likewise, the manual suggests that a data communication channel must be establish between the Operating Agencies and the Central Laboratory to report equipment malfunctions, data problems or any other issue that require central assistance (Airzone One and Bootsma 2011).

As stated by Heeks (2008), one of the main reasons behind the failure of early ITC4D projects was the lack of understanding of the current environment. Taking into account the different locations of the EADN sites and their data transfer and ubiquity capabilities, mobile phones represent a good choice for establishing a data between the site operators and the laboratory. However, there is one fact that must be taken into account about mobile phones in Africa. Only 17% of the African mobile phones are smartphones (Blycroft 2012). It can be assumed that any data transfer solution involving mobile phones in Africa will have to deal with handset that only offer basic speech and text capabilities.

The Unstructured Service Supplementary Data, or USSD, is a messaging service used in a Global System for Mobile Communications (GSM) network. GSM is the dominating mobile technology in Africa (Blycroft 2012). Similar to SMS, USSD transmit data messages using the signaling channel, so no data plan is required to access it. More importantly, USSD works on any kind of GSM mobile phones, from old handsets to smartphones. Unlike SMS that uses a store-and-forward transaction model, USSD provides session-based connections, a characteristic that makes it faster. As defined by Sanganagouda (2011) "USSD is as similar to speaking to someone on the phone as SMS is to sending a letter".

USSD applications are made of menus that users can browse using their phone to upload requested information to a remote server. These applications are accessed by dialing a number that starts with an asterisk (*), then a combinations of numerals and asterisks and ends with a pound symbol (#). Common users of USSD worldwide are mobile pre-paid subscribers that use it to check their balance or top up their accounts. In Africa, USSD is broadly used for other kind of applications like mobile banking, or to access useful information in rural areas like agricultural and live stock prices (Ekine 2010). Figure 10.6 shows a general diagram of a USSD application interacting directly with the EADN Central Laboratory.

Through the implementation of a simple menu base application sites operators could use their mobile phones to upload a sample SHF directly into the EADN Central Laboratory database without any intervention from the central lab staff. More importantly, because USSD offers network initiated messages, the central laboratory staff could use the USSD channel to send real time notifications to site operators, or even request them to upload more information as needed.

On the one hand, several are the benefits of incorporating USSD in the EADN; it's a commonly used technology in Africa so Site Operators might already be familiar to it, USSD prices are normally cheaper than SMS prices or, in some cases, free. Finally, as suggested by Ekine (2010), some Mobile Network Operators (MNO) keep USSD ports open for roaming traffic, so

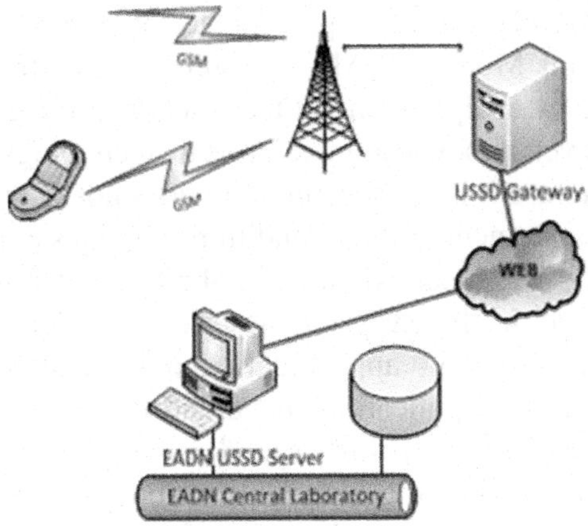

Figure 10.6 EADN USSD application

technically the USSD application could be accessed without problems from all the different EADN countries.

On the other hand, the development of any type of USSD application requires interaction and collaboration with the MNO, because they are the ones who manage or lease the USSD gateway from a third party owner. As a result, the development of the USSD application could be complex because of the different parties involved.

10.6.3 Educational Layer

The EADN represent a multinational effort aimed to understand the sources of nutrients deposition in the AGL. In the end, the information generated by the EADN will be input to policy makers while design and implementing environmental protections plans such as the Phase 2 of the Lake Victoria Management Plan.

As it has been mentioned, biomass burning related to human activities is expected to be one of the major reasons behind nutrients emissions. Consequently, reducing nutrients emissions, as any other kind of environmental concern, must heavily rely on education strategies.

Accessibility is one of the major challenges Africa faces. Lack of proper transportation is one the problems people in Africa have to deal with every day, a situation that gets worse in rural communities. According to the World Bank, by 2010, Kenya, Tanzania and Ghana had a percentage of paved roads of 14%, 15% and 13%, respectively. Actually, the paved roads percentage all across Sub Saharan Africa was of 16.3% by that year. Other basic infrastructure like electricity or sewage also have low percentages or accessibility. Through the Technology Task Fit Model (TTM), researchers have tried to explain technology acceptance by analyzing the fit that such technology represent in users everyday tasks. Even if they recognize a certain technology as advanced, users may decide not to use it if such technology does not fit their needs and make a significant improvement to their conditions (Junglas et al. 2008).

In the same way the TTM can explain why the telecenter approach used by many ICT4D projects in the early nineties were a failure (graffiti with the phrase "jobs not computers" appeared in India), it also explains the success mobile phone based applications have had in Africa. When taking into account the amount of time people spend moving around because of poor transportation systems, allowing them to have money transfers/payments, agricultural pricing information or health advices from their mobile phone, it is clear that such technology represents more than a communication channel in Africa, it represents a life changer.

Simple SMS-based applications have made a huge impact in all kind of sectors in Africa. In 2007 Safaricom launched M-Pesa in Kenya, a mobile applications intended to offer financial services to the unbanked[5] sector. Prior to the launch of M-Pesa, only 18.9% of Kenya's population had access to any kind of financial services. As a proof of M-Pesa's success, by 2011 70% of Kenyans used M-Pesa as a money transfer service (UNCTAD, 2012). In more socially oriented applications, SMS in Africa is being seen as major tool for activism. Fahamu, and African NGO with offices in Kenya, South Africa and Senegal used mobiles phones to promote the Protocol on the Rights of Women in Africa. A SMS alert service was set up so that users can sign up and receive updates of the progress of the campaign. This public awareness effort was key to the ratification of the Protocol by the Afican Union (Ekine 2010). The implementation of an environmental SMS-based Government to Citizen Application (G2C) as part of the EADN represents the following benefits:

- By means of using the MNOs' infrastructure as the communication channel instead of more traditional approaches to get in touch with the population, economic saving can be achieved.
- Based on mobile phones ubiquity, government agencies will be able to get in touch with communities in which normally it would be hard to access to.

[5]In the literature, the term unbanked refers to the population, mainly low income, without access to a bank account and the associated services.

- Launch awareness campaigns about increasing agricultural efficiency or education around non-wood fuel sources of energy, both activities related to biomass burning.
- Promotion of new economic activities that could reduce the overall regional dependency on agriculture.

10.6.4 Monitoring Layer

Just like the Educational layer proposes a G2C application to implement a top to bottom strategy for creating awareness regarding biomass burning, the first part of the monitoring layer describes how mobile text messaging can be used to develop a bottom to top plan.

Advocacy has found an ally in mobile phones. Kubatana, an information and civic activism platform was founded in Zimbabwe in 2001 as a way to countercheck the repressive policies implemented by the government against free speech and independent press. Even though it uses different media like web pages, email or print publications, SMS offer Kubatana the best way to allow their subscribers to share their thoughts and feedbacks about the platform (Ekine 2010).

In the same way that the Kubatana platform did, the EADN can take advantage of the direct link mobile phones offer. As part of the policy making process, evaluating the impact of the mobile educational layer is something that must be addressed. Through a Citizen to Government (C2G) mobile app consisting of a SMS-based inputs server, the EADN staff can receive notifications from local civil servants, environmentalists or general public members regarding non sustainable practices detected in their communities. When compared the inputs received with the awareness campaigns launched by the education layer, an analysis of the effectiveness of such campaigns can be achieved and, if needed, recommendations towards improving their impact can be generated.

The second part of the monitoring layer incorporates the notion of Big Data for Development. Created by the computing world, the term Big Data refers to "an umbrella term for the explosion in the quantity and diversity of high frequency digital data" (GlobalPulse, 2012). Sources of this digital data could be social media like Twitter of Facebook, web browsers inquiries, mobiles phones call records or wireless sensor networks. The private sector has had an interest in Big Data analysis like, for example, identify customer trends or marketing strategies planning.

In 2009, the Global Pulse Initiative was launched by the United Nations to explore what Big Data can do for Development. More specifically, the Global Pulse Initiative does research around the idea of how the analysis of these huge quantities of real time data can help decision makers to track plans' implementation progress, improve social protection or understand adjustments policies need (Global Pulse 2012). As expected, one of the challenges Big Data represent is the intensive use of data mining and mathematical analysis techniques to create "data of our data", a step needed to correctly understand the huge amount of information a typical Big Data information set represents.

In the Developing world, lack of basic infrastructure and mobile phones ubiquity potential are reasons behind the innovative ways people is using this technology. Access to medical or pricing information are just examples of how mobiles phones are becoming more a more an important part of developing countries.

A Call Detail Record (CDR) is a record generated each time a mobile phone is used to make a phone call or send an SMS. CDRs are key features of a mobile network management. Things like generating phone bills could not be possible without the use of CDRs. The information contained in a CDR is refer to as metadata, a description of a mobile session. A CDR does not include information about the content of the session. Examples of the information a CDR contains are: Sending and receiving mobile phones, type of mobile exchange (voice call, SMS, etc.), call duration, timestamp or antenna codes (Clayton 2001). As expected, access to CDRs requires interaction with MNOs and issues like privacy must be taken into account when dealing with this kind of information. Recently, the Computer Social Science, an emerging multidisciplinary research field, became interested in using CDRs to understand human behavior. For example; Gonzalez and others (2008), used CDRs to trace people's movements.

CDRs analysis represents a valuable source of knowledge to the EADN. Just in the same way the proposed telemetry stations will try to understand the long range transport of nutrients before they are deposited in the AGL, mobile phones, and their associated CDRs, will trace human behavior all across the EADN region. Examples of the types of analysis the monitoring layer will perform are:

- Identifying human density around the lakeshore areas: Using CDRs' antenna codes and the geographical position of the associated cellular towers, patterns of how the people moves in and out the lakeshore areas can be established.

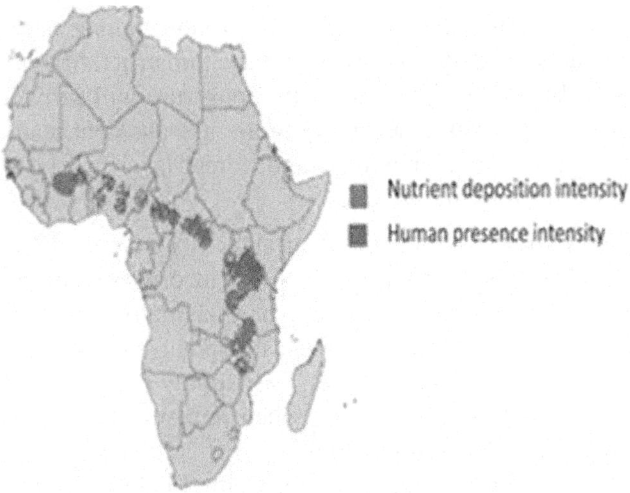

Figure 10.7 EADN correlated analysis

Identifying work related activities: As it has been mentioned, agriculture represents the major labor force occupation in the EADN region. It could also be one of the main reasons behind biomass burning. Taking into account how much mobile phones are used in work related activities, the analysis of the CDRs on work hours will help to understand how much agricultural activity was done in a certain area in a specific period of time.

Of course, the huge potential of this type of Big Data analysis will occur when correlated with the data obtained from the EADN telemetry stations. Comparing the nutrients pathways generated by the stations measurements with the human traces obtained from the CDR analysis will provide a more holistic understanding to nutrients emissions and human behavior. Figure 10.7 gives a graphic example of how such correlated analysis might look like.

10.7 Recommendations and Conclusion

Mobile phones and their impact in Africa represent a huge opportunity for environmental protection. In the past, the lack of data has been one of the major challenge science faced when trying to propose solutions to the emerging problems of the African Great Lakes. The Equatorial Africa Deposition Network is the first project aimed at identifying long range sources of nutrients and their transport patterns. However, because of the huge dependence and interaction the Lakes have with their riparian countries, human behavior must be included to have a more holistic approach. Using mobile phones as a communication channel and also as a way to trace human activity will enrich the data generated by the EADN and, in the long run, will provide better evidence to policy makers while designing future management plans for the African Great Lakes. For implementing the different layers of the mEF the following recommendations are provided:

- Following the policy of having one central facility to assure high quality analysis, a Mobile Data Laboratory should be incorporated to the EADN organization. The Mobile Lab should have a multidisciplinary team with experts with telecommunications, computer and social sciences background. The lab business model must follow the one used by the United Nations Global Pulse agency when implementing its Big Data labs located in New York City and Jakarta.
- The Mobile Lab should team up with the Central Laboratory to develop and support the USSD application Sites Operators will use to generate the real time Sample History Forms.
- Because it is the MNO with the biggest presence in the EADN region, MTN is recommended as the one hosting the USSD application.
- Through UNEP, the EADN should interact with the United Nations Global Pulse when developing the Big Data methodologies, with special emphasis in securing mobile phone users privacy.
- Supported by the African Telecommunications Union, a close relationship with the different Mobile Network Operators should be established. Projects like 2012 "Data for Development" Challenge realized by Cote d'Ivoire MNO, Orange, should be used as an example of how anonymous CDRs can be incorporated into the EADN while offering marketing benefits to the MNO's. Once implemented, the following recommendations are provided towards using the mEF.
- The EADN Regional Steering Committee should consider the mEF results as inputs of the policy development process.
- Through the SMS-based channel, the EADN Regional Steering Committee must implement educational campaigns aimed at changing unsustainable practices detected by the mEF data inputs.
- When doing a final evaluation of the EADN results and proposing inputs to management plans regarding the AGL, the nutrients measurements obtained should be correlated with the information generated by the mEF.

Acknowledgement

Thanks to the United Nations University Institute for Water, Environment and Health (UNU-INWEH) for facilitating the information related to the EADN definition and operation through its Water without Borders program.

11

Sustainable Energy Generation for ICT Development in Sub-Saharan Africa

Kenneth K. Tsivor
CMI, Department of Electronic Systems,
Aalborg University Copenhagen
Denmark
kents@cmi.aau.dk

11.1 Introduction - Telecom and ICT Growth in Sub – Saharan Africa

Telecom Industry has for some time now been playing central role in most economic and industrial development of developed countries. Following technological innovations, information communication technologies (ICT) have been embraced by many technology fanatics that it can play a significant role in shaping economic growth and improve gross domestic product (GDP). The banking and financial systems have benefited from the advancement in ICT through instant trading in the stock market, worldwide automatic teller machines (ATM) and the timely authorization of credit card transactions. It is also seen as having developmental role and the ability to expand capacity for employment, improve productivity, organizational restructuring, democratic participation of citizens as well as reduce poverty (Kozma, 2005). There is evidence that ICT is facilitating economic growth through increasing efficiency in OECD countries (OECD, 2004). It has contributed to the real gross domestic product (GDP) development and is fueling overall economic growth of many countries. Within the European Union (EU), ICT is considered an engine of growth since it has contributed to 5.6 percent of GDP between 2003 and 2005 (Forneld, Delaunay, & Elixmann, 2008). According to ITU report (2010), the percentage of people with access to communication networks is spreading very fast across the world. It has been extensively used in healthcare delivery sector, education and social integration of individuals with diverse cultural backgrounds and abilities.

In spite of this improvement, Sub - Saharan Africa countries still remains the least developed in terms of mobile telephony service in the world because the actual number of individuals without access to facilities is enormous. These demonstrated that the region still has large market for telecom operators.

Reliable energy is currently a key sustainability focus of ICT development in Sub – Saharan region. It has been demonstrated in several studies that the appropriate use of electrical energy has some level of impact on social and economic development of citizens in developing countries (United Nations, 2010) particularly in the case of many rural and semi–urban towns where electric power consumption contributes greatly to the operation and maintenance expenditure of telecom operators (close to 70%). This power consumption is relatively less in the urban cities (20 – 30%). This expenditure on electric energy is due to the lack of grid availability and reliability of supply and creates an impediment for telecom and ICT development, particularly in the rural areas where expansion in teledensity stands at 149 percent. It is estimated that in Sub – Saharan Africa, almost 70 percent of telecom base transceiver stations are located in areas with more than six hours of grid outage and 83 percent are located in off grid areas (Ogunlade Davidson and Standford A. Mwakasonda, 2004). Areas that are connected to the national power grids often suffer outages, and these have forced many countries to undertake load-shedding arrangement, or roll blackouts, in which residents receive electricity on a schedule that ranges from every other day to once a week or less. This unreliable power supply has compelled telecom operators to deploy diesel engine generators to ensure a continuous power supply for their equipment.

Currently, electricity consumption in the telecommunication and ICT sector across the world is estimated to be about 2 to 2.5 percent of the total electricity consumption of the world (Connected Urban Development, 2013) with a corresponding roughly about 2 percent carbon emission (Gartner, 2007). Even though, this may be small as compared with other industries, the fast rate of growth of the telecom and ICT industry will consequently lead to an increase in the energy consumption and it's multiplying environmental effects. However the relative benefits of mobile telecom when evaluating the environmental impact is significant. The use of mobile telecom and ICT for long distance communications and data transfer reduces the carbon emission which would have been generated from transportation.

Most developing countries are endowed with diverse forms of energy sources ranging from renewable sources such as solar, wind, hydro power to non-renewable petroleum sources. Renewable energy could have a significant role in providing reliable power to improve access to ICT in most developing countries at the same time reduce the emissions.

11.2 Telecom Architecture being used in Developing Countries (2G/3G Architecture)

Typical mobile network being used in developing countries are mostly of 2G/3G. These networks have the radio access network (RAN), including base transceiver station equipment (being BTS, MSC) and mobile radio core network with its mobile switching Centre. The radio access network houses data Centre. The data Centre is made up of IT equipment (servers, storage devices etc), the power and air conditioning infrastructure that supports IT and core network equipment. The power system within radio access network is responsible for providing constant backup power, regulate voltage, converts alternating current and direct current (AC/DC). It also serves the cooling equipment such as room air-conditioning units in the computer rooms.

By principle, electricity is supplied to an uninterrupted power supply (UPS) unit which serves as a backup battery bank for the entire system to ensure that all the units do not experiences any kind of power fluctuation and disruptions which could cause disruption to traffic or loss of valuable data for business.

Electricity is converted from AC to DC to charge the internal batteries within the UPS. The output of the batteries which is DC is then converted back to AC to serve the various equipment and gadgets. Within the UPS equipment, there are power units that control and convert the electricity to low voltage DC power which is then used by the internal components such as central processing units, memories, fans. The core network equipment and storage devices enabling the transmission of signal as well as data also depend on reliable electricity supply for their operations. Mobile operators in most developing countries use mainly 2G and 3G network; thus Global System for Mobile communication (GSM), Code - Division Multiple Access (CDMA) systems at the Base Transceiver Station (BTSs), Base Station Controller (BSCs) and Mobile Switching Centre (MSCs) for their operation. Since the BTSs, BSCs and MSCs are always in operation, a lot of energy is consumed. Modern IT equipment generates high intense heat; the necessary cooling to enhance effective operation ought to be provided. As the demand increases, the base stations of the mobile RAN are widely distributed throughout to match the demand and thus lead to an increase in the BTSs, BSCs and MSCs. This naturally leads to increase in energy consumption in the mobile telecom industry.

11.3 Energy Consumption of Telecom Base Station

The Base station power consumption mainly depends on the numbers of core network element, thus as the core network element increases, the power consumption also rises. Since base transceiver stations are always in operation, energy consumption increases accordingly to match the demand (Koutitas, 2010). A typical Radio Base Station requires about 10.3kW energy to produce 120W of radio transmission. This makes the Radio Base Station consumption accounts for nearly 90 percent of the entire consumption of a mobile network. A Radio Base Station comprises of Radio equipment, DC Power, Cooling, Radio Frequency (RF) load and Feeder (Gildert, 2006). Figure 11.1 shows an arrangement of a Base Station. The DC Power and the cooling system use 1170W and 2590 W respectively. While the power consumption of the Power Amplifier is about 4160W and the signal processing and control units consume 2190W. The consumption of the feeder and the Antenna is about 120W each. Figure 11.2 shows load distribution and the percentage of energy consumption.

From Figure 11.2, it is clear that the power consumption of the telecom base station is huge so there is the need to get the necessary power sources to meet this growth. In addition, the rising of oil price in the international market

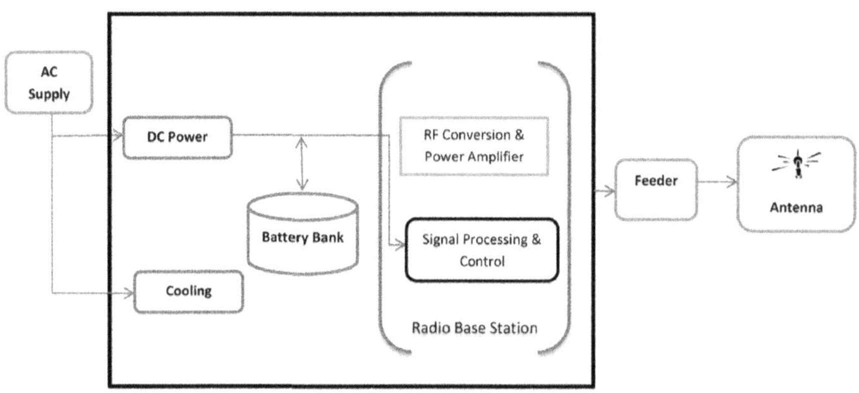

Figure 11.1 Base Station Arrangement

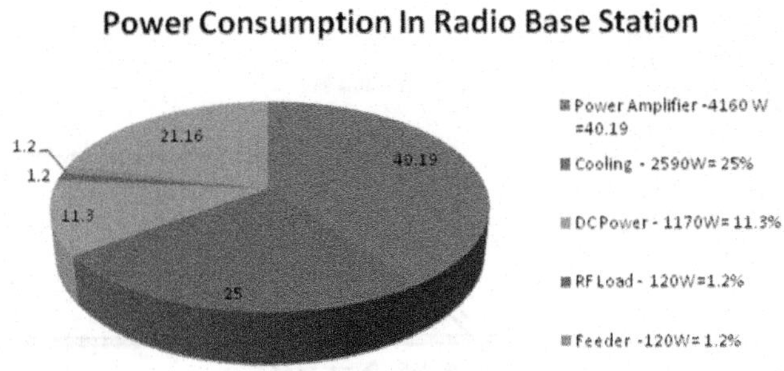

Figure 11.2 Power Consumption in Radio Base Station

Source: Emerson Electric Co.

makes energy costs keep rising leading to higher operation and maintenance expenses of the telecom operator. A single diesel engine generator (26kW) estimated to powered a typical GSM base station load of 18 kW, can consume around 57,634 litres of diesel per year (i.e. 6.58L/H) (Rebecca Mayer, 2007). It is therefore imperative for the telecom operators and the industry as a whole to seek an alternative source of energy that can improve network operation and reduce energy costs.

11.4 Overview of Solar Photovoltaic Technology Renewable Energy

Solar Photovoltaic Technology is a form of renewable Energy whose resourcesare regenerative. Energy generated from renewable sources do not emit greenhouse gases, and has been accepted technically as a technology that produces reliable power and is competitive to other forms of energy generation. Solar photovoltaic technology uses the light (photons) from the sun to produce DC electricity. As shown in Figure 11.3 below, a photovoltaic cell is a light-sensitive semiconductor device which, when exposed to sunlight, releases electrons to produce DC current. The technology comprises of solar photovoltaic cells which are arranged in series and parallel to form solar photovoltaic power system. The cells circuitry are laminated and packaged to prevent moisture from entering and avoid corrosion. A number of panels are connected in series and are termed as a solar photovoltaic string. Solar photovoltaic arrays are a group of strings which form the complete power generation unit. Figure 11.3 illustrates a solar cell, module, and array and string structure.

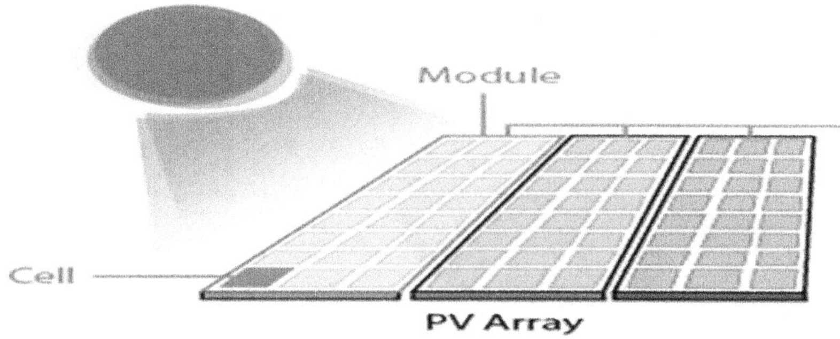

Figure 11.3 Electricity generation in a solar photovoltaic cell

The system also has charge controller and battery bank. The charge controller in a photovoltaic power system regulates the voltage and current output from the solar panels as required by the battery and the load. It also keeps the batteries protected from overcharging and discharging. While the battery bank stores power against bad weather and during non – sunshine hours. Battery capacity is measured in Ampere-hours (Ah) at a constant discharge rate. Variety of batteries can be used in solar photovoltaic configurations. However, the commonly used types are the Lead-acid and valve-regulated lead-acid (VRLA) gel batteries.

11.4.1 Solar Photovoltaic Applications

Solar photovoltaic technology can either be used as a stand-alone, grid-connected or hybrid solution. The stand-alone type of application requires back-up energy storage to ensure sufficient power supply whenever the sunshine is unavailable. The grid connected has energy fed back from the photovoltaic modules to the grid and the hybrid is the combination of photovoltaic arrays and other energy sources such as hybrid with wind turbines, biomass power, and fuel cell and diesel generators

11.4.2 Considerations for Photovoltaic Applications

Geographic location, sunshine duration, daily energy incidents and solar power density influences the application of photovoltaic system. Considering a typical tropical sub-Saharan African country such as Ghana, Figure 11.4 shows the solar power density across the country mapping the performance and deployment feasibility of solar photovoltaic solution. Table 11.1 provides facts on solar radiation across the Ghana.

Table 11.1 Solar radiation in Ghana

Parameters	Region	Availability
Daily Average solar energy incident	North	7 – 7.5 kWh/m2/day
	Middle Belt	6.5 – 7 kWh/m2/day
	Eastern sector	5–5.5 kWh/m2/day
	Western and Central	4.5 – 5 kWh/m2/day
Duration of quality sunshine/day		Approximately 5 – 6.5 hours
Number of days of quality sunshine		330 days in a year

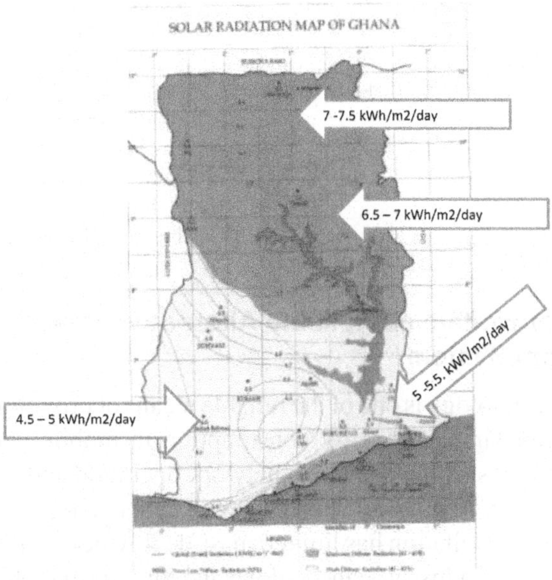

Figure 11.4 Solar power density in Ghana

Source: UNEP

11.4.3 Challenges and Advantages Solar Photovoltaic Technology

Currently, the dependency on sunshine availability and configuration of storage capacity has contributed to the high initial investment in solar photovoltaic technology. This has become limitation for its acceptance for many applications. Table 11.2 provides a broad overview of some of the basic advantages and challenges of solar photovoltaic applications.

Table 11.2 Advantages and challenges of solar photovoltaic technology

Parameters	Advantages	Challenges
CAPEX	Now low due to mass production and technology innovations	Requires high storage capacity, hence many batteries cost increases the CAPEX
OPEX	No fuel needed	Regular solar panel cleaning is necessary to maintain optimum efficiency
Number of days of sunshine	330 days in a year	Forest areas of Ghana have lot of rainfall therefore less sunshine duration
Storage	Enough sunshine to charge batteries (4 – 7.5 kWh/m2/day)	High capacity batteries are needed in areas with less solar intensity
Emission	Zero	None
Configuration	Easily integrated into hybrid	Needs equipment automation for optimum solar photovoltaic usage due to intermittent sunshine availability. A high capacity leads to a high CAPEX investment

11.4.4 Solar Photovoltaic as Reliable Alternative Power for Telecom System

Solar photovoltaic technology is one of the reliable means of providing reliable power as well as reducing the energy consumption in mobile networks in order to minimize the costs incurred in the network operation and also reduce the carbon dioxide (CO2) emission. However, the dependency on sunshine, CAPEX and space for installation has limited mass deployment of the technology. Within the Telecom Industry, stand – alone and hybrid solar photovoltaic application has been identified as most appropriate. The application types were chosen based on the site load profile, grid outage scenarios, space availability at the site and other configuration aspects including average sunshineavailability throughout the trial and the power storage configuration for non-sunshine hours.

11.4.5 Solution Design Considerations for Africa (Ghana)

The use of solar photovoltaic technology as an alternative energy makes it possible for areas without grid power supply to enjoy the benefits of mobile communications networks Currently most developing countries have difficulties with electricity generation and distributions. Since most developing countries depend on oil and gas imports for their power generation, it imposes a heavy burden on their governments and therefore affects electric power generation and distribution. For years, mobile telecom operators depend on diesel generators in communities that do not have access to a reliable electricity grid for their operation. Diesel fuel is expensive to begin with, and its transportation and storage can increase its cost considerably, especially in remote areas. Mobile telecom operators in developing countries lament that diesel fuel and its associated costs increase their entire operational expenses (more than two-thirds). The situation is likely to get worsened in the near future as there is an increase in fossil fuel prices. Also, the use of diesel fuel (which is part of fossil fuels) causes environmental damage. These problems are not peculiar to developing regions. Research suggests that electricity costs will rise faster than the savings gained through more efficient equipment and networks.

Fortunately, the geographical location of most developing countries especially African countries gives them several advantages for extensive use of most renewable energy resources which appear to be clean and effective alternative energy source. The solution design is based on the availability of sunshine in particular geographic regions. Many African countries especially Ghana has average sun shine duration, suitable wind speeds and other natural resources abundantly which are perceived to play pivotal roles in the future energy generation for the Telecommunication and ICT industry.

Solar photovoltaic technology as an energysource needs the capacity to support the BTS load afterconsidering the losses of the battery, charge controller and otherauxiliary loads. To do this, the BTS site load profile is necessary for solar panel design. Also the site equipment efficiency, charge controller, battery and other loads efficiency influence the solar panel capacity. Statistical data shows that the daily average energy incident availability in Ghana ranges between 4 and 7 hours which impacts greatly on the output of each panel and this helps in determining the capacity at each site. Though efficiency system may vary, the efficiency of different panel sizes influences the total solution objective. Also, the number of panel required for the energy demand is determined by the panel efficiency. The battery configuration and charging current limitation of a given battery is fixed andbased on its specification. Battery capacity is designed accordingto the duration and availability of sunshine and charging currentlimitation, especially when solar is the only source of battery charging.

11.5 Case study: Solar Simulation & Results for Telecom Base Transceiver Station

By using Hybrid Optimization Model for Energy Renewable (HOMER) simulation, the studyused the meteorological data of Ghana (West Africa) for HOMER simulation to examine the viability technical feasibility of Renewable Energy (solar Photovoltaic) in a mobile network. The study considered the energy consumption for a typical 2G/3G GSM telephony base station site since this is the technology being used most of the developing countries. It also considers load between 0.7kW to 4kW with load demand of approximately 31kWh/d for a base station as average in developing countries. The study assumed the availability of reliable sunshine all year round. It is appropriate to exploit the use of hybrid power system in mobile base stations where the national grid is either not accessible or not reliable. From simulation results, the most technically and economically viable renewable energy configuration system consists of 3kW solar array, 12kW inverter, 56 pieces of 6V battery, 15kW diesel engine generator.

The system adopted in this study is a decentralized system which permits maximum use of solar PV that eliminates the resource fluctuation, reduces energy storage while increase the overall power output. It allows the outputs of both Solar PV systems through a DC – DC converter and output of the diesel engine generator (DEG) through AC – DC rectifier to provide the system load with some storage batteries banks as backup. The diesel engine generators (DEG) may be used during emergencies periods in order to increase the reliability of the telecom network. Electricity is generated as and when the sun rays hit the solar PV panel. The proposed PV system is a simple control which combine maximum power tracking control that makes use of diode characteristics or PV system that features output stability with input dc-dc converter capable of controlling and monitoring the current and voltage output of storage battery bank charging (Ashok, 2007), (Hirose & Hirofuni, 2012)

The relevant technical details from the simulation are shown in the schematic diagram in Figure 11.5. Table 11.3 show a summary of the life time costs of different configurations, as well as capital cost and the overtime costs of different configurations.

From Table 11.3, the selected configuration for a Base Station using minimal renewable energy meets the demand and keeps the diesel engine generator as a stand- by for bad weather is highlighted.

Table 11.3 Summary of life time costs in Ghana

Conf.	Initial Capital ($)	Operating cost/year ($)	NPC ($)	COE ($/kWh)	Diesel (L/Year)	Label (Hrs)
GBC	42,900	9,724	167,211	1.145	6,384	1,991
GBCW	72,900	10,248	203,910	1.396	3,552	1,176
GBCP	63,900	6,900	152,146	1.042	4,122	1,353
GBCWP	93,900	7,314	187,398	1.283	3,552	1,176

The final outcome has the following characteristics: NPC and COE were $ 152,146 and $ 1.042/kWh respectively, lower than any other configuration that could meet the same load demand. The total annual electricity production is 14,885 units (kWh), and the annual operating cost is $ 6,900. Compared to the diesel-only systems there are a fuel savings of $ 21,077 thus 4,122 litres of diesel per year. An annual carbon dioxide emission reduction of 10,854kg was achieved. An annual renewable energy contribution of 33 percent all came from wind resource. Figure 11.6 shows the monthly average electric output from the three sources considered.

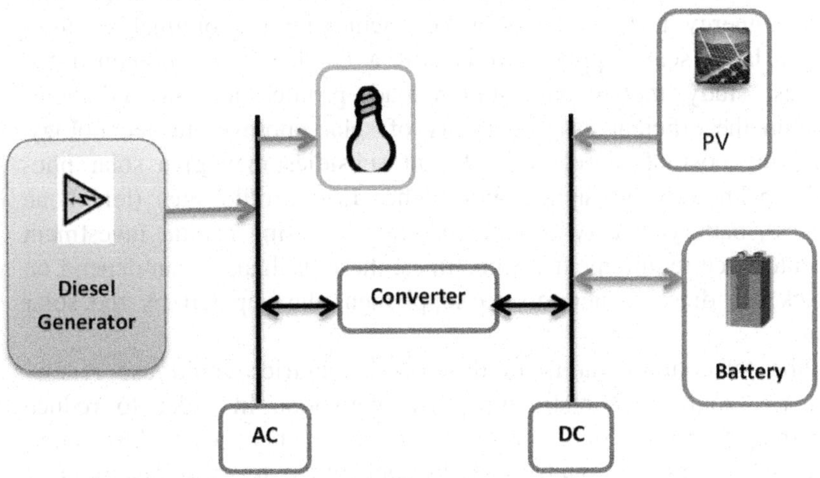

Figure 11.5 Reliable power systems for mobile telecom base stations

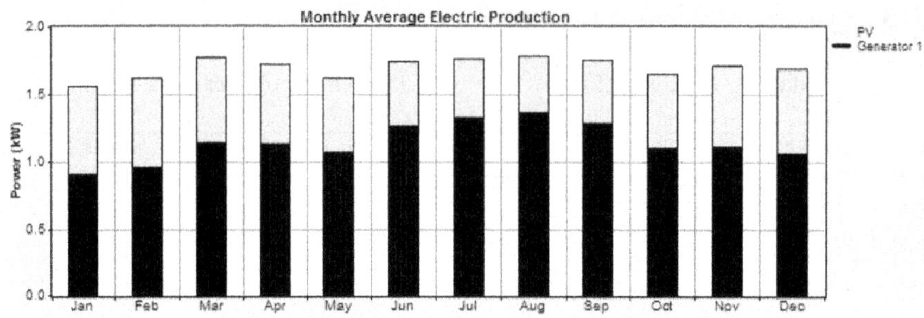

Figure 11.6 Monthly average elastic productions from the three sources considered

11.6 Conclusions: Future of Solar Photovoltaic Technology for Telecom

The need for reliable power will continue to rise as the demand for ICT innovation keeps evolving in developing countries. The issue of renewable energy is not limited only to developing countries. In telecom and ICT industry, solar photovoltaic technology has a greater advantage and has enjoyed a better rate of application to date as compared with other renewable energy technology. However, achieving the optimal configuration for large scale application is still a challenge. As demonstrated in the case study, mobile base station loads profiles are low and therefore within the efficient usage capacity of solar photovoltaic technology. The growing cost of diesel and relevant subsidies may give solar photovoltaic technology advantage. For higher load profile sites (telephone exchanges), batterycapacity is high thereby increasing capital investment and maintenance requirements. Sometimes, these exchanges can depend on (fall back) on diesel generators to supplement the gap left by the solar solution.

Mobile telecom operators in developed countries could use renewable energy sources to restructure their networks, in order to reduce their carbon footprint and control the rising energy costs. Also Government of developing countries should encourage telecom operators to use solar photovoltaic and other renewable energy sources at their base stations. This can be an excellent way of avoiding the intermittent power outages and could help keep the success so far derived from the ICT applications. This is because renewable energy technologies are economically and technically feasible in most developing countries especially Africa.

Acknowledgement

This works was supported by Ghana Telecom University College. The author is grateful to the staff of Centre for Communication, Media and Information Technologies (CMI), especially Prof. Knud Erik Skouby.

12

The Role of Mobile Telephony to the Deployment of Intelligent Transport Systems in Africa

Daniel Michael Okwabi Adjin
CMI Aalborg University
Copenhagen, Denmark
& GTUC Ghana
adjin@cmi.aau.dk

12.1 Introduction

12.1.1 Preamble

Invariably, information, knowledge and technology are accepted as the key driving forcesbehind socio-economic development and technological advancement of societies and nations globally. However, the manner in which information is handled and managed to support socio-economic activities, particularly in the transportation sectors of many African countries leaves much to be desired. ICTs are spectacularly changing the life styles and living standards of many people in the developed world; primarily, in Europe, USA and some leading countries in Asia (e.g., Japan, South Korea, China, etc.). Indisputably, the trend is not the same in Africa. The capabilities of African countries to accelerate their socioeconomic development process, and more importantly, to gain competitive advantages and improvements reckon deeply on the degree to which these nations can develop ICT, and for that matter ITS technologies into their traditional transportation sectors in diverse practices and levels.

Arguably, many African countries have not yet developed ICT into their traditional transportation sectors; thus, transportation sectors in these nations do not have most of the technological advancements required in their transportation industries; hence, it is for this reason that this chapter is aimed at analyzingthe prominence of the transportation sectors in Africa.

12.1.2 Chapter Objectives

The prime focus of the chapter is to analyze attempts made in Africa to develop Intelligent Transportation Systems (ITS) in their Conventional Transportation Sectors (CTS). The chapter discusses what the case was (i.e., the status of the transportation sectors in Africa) before the adoption of ITS technologies, and what the current state of ITS development in Africa is today. Precisely, the discussion is based on country case studies; and for the purpose of this chapter, Ghana is the country case with the aim of ascertaining how far has Africa adopted, developed and deployed ITS technologies into their existing transportation sectors. Adding to this chapter is to also highlight the necessary further steps that African nations need to take to be able to develop ITS technologies and solutions rapidly into their extant transportation sectors, in order to improve and advance the socio-economic activities and living standards of the people of Africa.

12.1.3 ITS Defined and Explained

Prior to the analysis of the status of transportation sectors in Africa before the adoption of ITS solutions, it is very imperative to define and explain the term ITS as applied in the fields of ICT and mobile development in general. By definition, "the telematics that apply new information processing and communications technologies to address vehicular traffic and transportation problems for the sole benefits of humanity" (Pamadas, Nallapernunal, Mualidharan, & Ravikumar, 2010) or society and public good at large, is appropriately referred to as Intelligent Transportation Systems (ITS) technologies or ITS solutions. The ITS Standards Bodies' further define ITS as "a Technology that combines ICT to transport infrastructure and vehicles in an effort to manage factors that typically are at odds with each other" (Eichler , 2007); these includes vehicles, road networks and human (people), see figure12.1, in order to improve road safety, live saving rates, to reduce vehicle tear & wear, to mitigate fuel consumption & carbon emission, and to diminish travelling & waiting times. By this definition, ITS plays essential role, as it "can provide innovative solutions aimed at: realizing a vision of paperless information flow (e.g., e-Freight), increasing efficiency of vehicular traffic management and making more use of technologies for tracking and tracing vehicles".

To explain briefly, these ITS are a hybrid of ICT enabled machine-to-machine communication, connecting vehicles and transport infrastructures (including road signs, cameras & sensors, etc.) and person-to-machine communication. The prime conceptsunderpinning the model of ITS technologies that constitute their ubiquitous potentials are: Advanced Traffic Management Systems (ATMS); Advanced Traveler Information Systems (ATIS); Advanced Public Transport systems (APTS); Commercial Vehicle Operations (CVO), Advanced Vehicle & Highway Systems (AVHS), Electronic Toll Collection (ETC), Collision Avoidance, Front End Collision Avoidance (FECA), Etc.(Kumarm & et al, 2009).

Technically, ITS is a family of services and technologies that apply ICT in the road and transportation sectors (Luc, 2005). It is the application of IT (computers, sensors, wireless communications and databases) to solve the problems of road transportation. It encompasses a broad range of wireless and wire-line communication-based information, control and electronic technologies. The underlying intents of ITS technologies are the development, deployment and operation of Advanced Road Traffic Management Systems (ATMS), Traveler Information Systems (TIS), Vehicle & Traffic Control Systems, Commercial Vehicle Operations, and lastly, Public & Rural Transportation Systems (PRTS).

ITS is the application of Information and Communication Technologies (ICT) to surface transportation systems, which maps relationships between human (people, users, organizations, etc.) and material things (nonhuman, e.g., objects, structures, etc.), which are called "Actors", as posits by the Actor Network theory (ANT), (Lewis, & Townson, 2004).

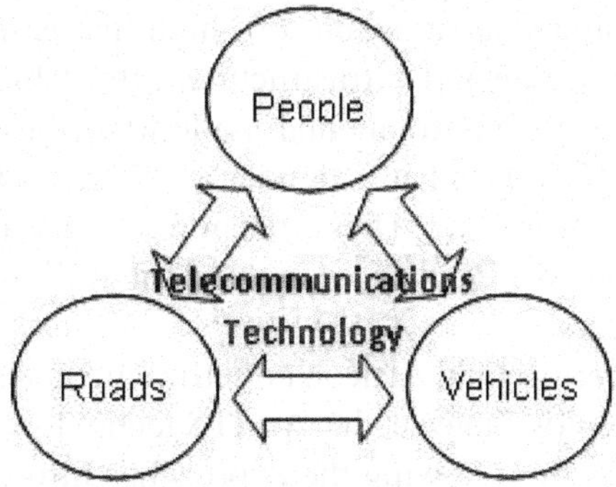

Figure 12.1 Illustration of ITS with reference to ANT

The figure above illustrates the link or relationship between the elements that traditionally constitute the design of ITS Technologies. These elements are composed of: People, i.e., Human Institutions; Roads (nonhuman), i.e., Transport Infrastructures and Vehicles (nonhuman), i.e., the Objects; they are associated or linked by Telecommunications Technology, integrated with IT Networks (Computers, Servers, Routers, Switches, Etc.). The key objective of ITS are to build integrated systems for People,Roads and Vehiclesin order to resolve problems of road transportation, including vehicular traffic congestions, accidents and environmental damages.

12.2 Status of the Transportation Sectors in Africa Pre-Its Development

12.2.1 General Status of the Transportation Sectors in African Countries

Globally, Transportation predicaments have major impacts on the quality of life of all citizens in all nations, their environments and economies, and African countries are not exempted from these dilemmas. Recent studies into ITS development and deployment activities around the globe point to the fact that, these impacts are more prevalent in developing countries, mainly those in Africa, thus these countries are often at disadvantage, compared to developed countries, regarding the development and deployment of ICT infrastructure to address and enhance technological advancement of transportation networks in many African countries.

For these reasons, transportation authorities in these countries are overwhelmingly encountering numerous challenges centered on worsening road traffic congestions as a result of: high increase in urbanization, lack of sophisticated transport infrastructure, affordability constraints, increasingly contributing to environmental pollutions through the emission of carbon dioxide (CO_2) and its attendant - climate change (climate degradation) and global warming. These phenomena are responsible for growing transport-user needs, wants & demands for ICT in transportation sectors.

Another serious concern is that, African countries are not able to catch up rapidly to develop and deploy State-of-the-Art ITS technologies through ICT to address these problems by following the approaches through which these same problems are being addressed in countries in the developed world.

12.2.2 Status of the Transportation Sector in Ghana

Transportation is a major source of sustenance for the Ghanaian economy. Despite its importance, however, the sector is faced with several problems, such as deplorable road conditions, poor vehicular maintenance, and poor law enforcement, all of which have contributed to very high vehicular crash rates in Ghana. The deprived road network is mostly seen in the disparity between rural and urban areas, where almost all the regional capitals and most of the district capitals have accessible roads while most rural areas have deplorable road conditions. The end result is that the produce, in particular, major exportable perishable commodities on which Ghana's economy depends, are subject to decay/rotten in the inaccessible rural areas, and create disincentives for farmers to produce more food to feed Ghanaian people. Not only are the transport networks worse off, but there also exist inequalities in motorable and accessible roads in the country, attributable mainly to economic resource none-availability in different areas. Lack of accessibility to critical destinations such as jobs, schools, markets, and health care centers has affected development activities in inaccessible areas. Besides, many of the roads have inadequate signs or pavement that is not equipped to handle the traffic. The country also lacks an effective and efficient public transportation system.

The Major Transportation Problems facing Ghana as a nation has to do with the deteriorated status of the transportation sector as a whole. The national economic products & commodities in Ghana, such as: agricultural, petroleum and other minerals are surface road-hauled for export to sustain her economy. Unfortunately, there exist disturbing transportation harms bedeviling Ghana as a country. Critical transportation barriers facing Ghana embrace the following (Adjin & Tadayoni, 2011): "Lack of modern and efficient vehicle tracking systems; In-effective communication between stakeholders due to lack of modern ICT in the national transportation network; inappropriate use of transportation resources by drivers and other personnel; and, more seriously, manual tracking of vehicles based on physical escort system which is unreliable, costly and outdated; loss of vital production times and revenues to stakeholders in the transportation industry, due to the absence of ITS applications in the conventional transport infrastructure in Ghana; lack of co-ordination, effective scientific monitoring and evaluation of road safety planning strategies and programmes; inability to set ITS standards for vehicle tracking and road safety; non-existence of research into the field of ITS studies in Ghana, etc."(Nii Darko & Janusz, 2010).

Table 12.1 Key issues in ghana's transportation sector

Issue	Sector Attribute Case
Effectiveness/Efficiency	Severe congestion in Accra Metropolitan Area. Lack of efficient Public Transportation Systems and poor infrastructure
Economic	Low income status, Significant % of agricultural exportable produce decay due to poor road access: disincentive to farmers to produce.
Social Equity	Wide disparities in road access infrastructure condition in urban and rural areas.
Environmental	Rapidly increasing trends in carbon dioxide, nitrogen dioxides, and non-methane VOCs in Accra
Decision Making	No formal mandate found for integrated land use/transportation planning
Other Related Issues	Poor Infrastructure Conditions, Poor Law Enforcement, Lack of emergency medical services, Poor land use planning in metro areas

This transportation situation has arisen in Ghana in view of the fact that, over the years, the few private institutions undertaking transportation activities in Ghana have failed to deploy ITS applications to ensure the installation of efficient ICT systems, hence, the prevailing alarming level of transportation problems in Ghana. These barriers can be removed/resolved by applying efficient ITS tracking systems employing Satellite Communication Systems, GPS, GSM Networks and Traffic Information Management Systems (TIMS) with efficient ICT data-centre networking.

The Ministry of Roads and Transport is responsible for the development and maintenance of transport infrastructure and the provision of transport services for all modes of transport in Ghana. The Ghana Roads Sector Development Program aims at achieving sustainable improvement in the supply and performance of roads as well as road transport services in a regionally equitable manner (Jeon, Amekudzi, & Vanegas, 2006).

Some issues of mass concern in the transportation sector of Ghana are summarized and tabulated in Table 12.1 below.

12.3 The Current Prominence of ITS Development in Africa

Interestingly enough, incorporating ICT into the existing transportation sectors in Africa has not been an easy economic task for many African governments. The general status of the transportation sectors in Africa is alarmingly deteriorating from city to city and from country to country, due to rapid and uncontrollable urbanization and motorization. The capacities of the national transportation networks in many African countries are overwhelmed with vehicular traffic volumes. These phenomena are occurring as a result of the fact that, the transportation infrastructure in many urban areas lack basic ITS applications. Further to that, there are inadequate ITS installations, together with the lack of legal transportation regulatory frameworks in many regional centres, as well as in some national capitals, for instance in Accra, Ghana.

12.3.1 Status of ITS Development in Ghana

In Ghana, there is a widespread concern for the environment, lifestyles are changing and information communication technology (ICT) and telecommunications are beginning to present a real alternative to travel, especially for urban transport. The ICT employment in the transport sector has been a more difficult area for Ghana which holds the key to modern transport direction as its involving branches have not seen an equally fast development to support the transport sector (Ghana Web, 1994).

Undoubtedly, it is not quite easy to identify any good demonstrations concerning development of ITS technologies in Ghana. Currently, the capacity of the national road infrastructure is approaching its bounds due to uncontrollable, and perhaps, unpredictable transportation needs, and utilization of private automobiles rather than communal transportation services. Conventionally, vehicular traffic control and management in Ghana is being done physically (by human interventions), sustained by a bit of "actuated-electronic traffic light signals"; however, no ICT and vehicle location devices (GPS) are integrated, meaning that, ITS applications are regrettably not incorporated. For instance, "Electronic Toll Collection points, Traffic Information System (TIS), Variable Message Signals, Fleet Management Systems, Traffic Management Centres and Transport Terminal Data Centres" are not available.

Just like many African nations, Ghana faces speedy motorization and uncontrollable vehicular utilization without corresponding development of modern transportation infrastructure. Vehicular Mobility and access to major road networks are blocked more often than not due to perennial vehicular traffic congestions, principally, during commuting peak hours. Distressingly, this calamity is worsened during festivities (religious & traditional), typically, during Easter & Christmas celebrations. Fascinatingly, metropolitan developers and state institutions responsible for the transportation sector of the national economy have come to realize the exigency to develop ITS technologies to address the awesome transportation problems facing Ghana before commercial cities and major road networks in Ghana are halted, due to overwhelming transportation deficiencies in the country (Jafaru, 2012).

The single trait of ITS solutions deployed by operators and investors in the surface road haulage business in Ghana has to do with Intelligent Vehicle Tracking Technology (IVTT). However, to a very limited extent, IVTT is deployed by few private companies in Ghana; entangled with inefficient GSM network coverage, absence of transportation monitoring and management systems, ineffective control of human behaviors & attitudes. In view of this situation, Ghana must institute nationwide ITS development strategies, to embark on mass deployment of ITS technologies and solution to her transportation sector within the nearest and shortest possible time frame.

12.4 Guidepost to Efficient Development of ITS in Africa

12.4.1 General Guideposts

Generally, there is the need to establish regional cooperation and collaboration between African countries to enable them to mutually improve upon their transportation sectors. For this to happen, there must be no traditional, cultural, institutional and political influences on any ICT/ITS development plans and projects for the transportation sectors in any country in Africa. It should be the vision and mission of all central governments of African countries to develop and implement long term ICT for ITS Development plans to inject funds into ITS projects and the systematic and regular undertakings of ITS projects with technical assistance from external ITS experts and ITS researchers. Such efforts require the readiness of the central governments to financially support the development ICT into the national transport networks by applying the ubiquitous potentials of ICT for development (ICT4D).

Additional ICT patterns worth adopting and practicing in the transportation sectors of many African countries include: the establishment of public policies to promote public transport operations; enabling ICT for initial applications of Advanced Public Transportation Systems, these may be categorized as vehicle location and navigation technologies; communication technologies; ITS standards; dispatching software; electronic fare collection & traffic signal pre-emption. All these are possible with the application of ICT solutions.

Early enabling actions that could be taken to implement ICT solutions in the transportation sectors in Africa include: developing consensus among stakeholders (actors/actants); developing effective organizations for early and continuing ICT facilities and actions transportation infrastructures; running of pilot tests; lessons & precautions from previous experiences, including the need for high-quality information and services; user orientation and overall ICT for ITS architecture planning at national levels.

Learning from the rich experiences of world leaders in ICT for ITS technologies, countries with transitional economies in Africa will have to study, pursue and adopt a redefined and well-tailored sets of paths closely to develop ICT solutions into their conventional transportation sectors. The institutional preconditions for initial ICT for ITS applications, including the development of new institutional arrangements (e.g. co-operation between public and commercial transport operators) must be established by African governments to provide effective and efficient ICT leadership in the transportation sectors of their national economies.

The establishment of basic requirements; such as government subventions, technical supports, user adoption and preparation for ICT in transportation operations under new environments (e.g. integration of electronic fare systems, detection of gas emission levels, detection of falsified number plates, etc.), must be in place prior to the commencement of any ITS project. It is therefore, the mission of this chapter to identify factors that will address these issues that mitigate the development of ICT technologies into the transportation sectors of the economies of African nations.

Countries on the African continent with transitional economies can take lessons from the great proficiencies and capabilities of the world's ITS leaders in a bid to study and embrace carefully the ensuing steps to develop ITS applications as posits the theory of modernization (Markus & Robey, 1988). The established prerequisites for preliminary ITS technologies, as well as the creation of novel influential agreements are: the instituting of Public Private Partnership (PPP), regulatory, monitoring & evaluating bodies by African governments, to deliver comprehensive and well organized ITS governance frameworks.

The formation of elementary necessities; e.g., "government subventions, technical supports, user acceptance and preparation for operation under new environments", must be in place preceding the initiation of all ITS activities in Africa as a continent.

There is the need for such institutions and agencies to have and retain extraordinary degrees of independence from the central governments in

Africa, without any forms of "political influences, traditional and cultural interferences" (Pinch & Bijker, 1992), (IBM, 2011). Adding to these, the following ITS solutions are worth adopting and implementing by African nations; these include: "the establishment of public policies to promote public transport operations; enabling technologies for initial applications of Advanced Public Transportation Systems (APTS), together with vehicle location and navigation technologies; communication technologies; ITS standards; dispatching software; electronic fare collection & traffic signal pre-emption". Prompt facilitating arrangements that must be put in place need to embody: "developing consensus building among stakeholders; developing effective organizations for early and continuing actions; the running of pilot tests; lessons & precautions from previous experiences, including the need for high-quality information and service; user orientation and overall national ITS planning".

One of the crucial determination processes in implementing ITS technologies in African countries is to first develop tailored ITS applications, and then to deploy appropriate ITS solutions, with the primary objectives of:- "Establishing ITS blueprints to meet national needs (and not individual needs as pertains now in many African countries); Systematic implementation of ITS projects; Drawing clear targets and measures for development of ITS technologies & standards that meet both local, regional and international technological advancement, requirements and market demands" (Markus & Robey, 1988).

In their attempts to react to the afore mentioned necessities of home-made ITS strategies and methodologies, transportation experts in Africa can no more be content with conventional methods of constructing fresh or extra transportation systems/facilities. The reason being that, these strategies require large and lands and other related resources, including huge financial obligations and multifaceted "regulatory and environmental planning processes to manage" (Jafaru, 2012).

To provide solutions, transportation entities worldwide are aggressively concentrating on the employment of "demand management with schemes like road user charging and information & customer management techniques including enhanced traveler information services". The new and extensive solicitation of Information Technology (IT) offers outstanding prospects to generate and drive novelties into the deployment of ITS solutions. The key strategies discussed above are what Africans must beckon on to deliver effectively and efficiently ITS solutions.

In addition to the general guideposts outlined above, reflecting on how ITS can be efficiently developed in Africa, this chapter further identifies tailored strategies that must be put in place, specifically by individual African countries to develop ITS solutions into their conventional transportation sectors. These steps are discussed in turn in the subsequence subsections that follow.

12.4.2 How ITS can be Efficiently Developed in Ghana

There is the need for Ghana to integrate land use and transportation decisions better to gain cutting edge control and management over vehicular traffic congestions, sprawls, and the associated air quality problems in the urban and metropolitan areas. The Ghana government's Vision 2020 is a road map to achieving middle-income status by the year 2020.

In the Vision 2020 framework, the fundamental policy objective of the transport sector is to establishan efficient, modally complementary, and integrated ITS transportation networks for the movement of people and goods at the least possible cost within the country.

Potential areas for improving the transportation sector in Ghana include, but not limited to: Improving roadway and pedestrian safety, improving emergency medical services and law enforcement; implementing effective and efficient public transportation systems; improving road accessibility especially for agricultural incentives; improving infrastructural condition particularly in rural areas; and improving land use and planning in urban and metropolitan areas (Jeon, Amekudzi, & Vanegas, 2006).

12.5 Conclusions

Unarguably, the glaring challenges of the transportation sectors in Africa today remains quite huge as comparedto the current level of continuous technological advancements in the transportation sectors in the developed nations. The solution to developing and improving transportation sectors in Africa, as well as in Ghana, is however not insurmountable if efforts are put into human, technical, capital and technological resources.

The transportation sectors are now broader aspects of many economies in Africa today, in the sense that several areas of human technology and institutions are needed to make the necessary developments and upgrading of the different transport systems and devices for all modes of transportation in Africa. The proposed guideposts discussed in the last section above will be good points of reference for the advancement of transportation sectors in the challenged transportation environments facing Africans. Mostly a measure of development recognized in many transportation sectors including Ghana focuses on the level of infrastructure available but this chapter clearly reveals that human capital and strategic transportation management policies must be boosted with efficient Research and Development (R&D) activities/projects in the transportation field to establish and maintain efficient transportation sectors in individual African countries.

Bibliography

[1] Nii Darko, Q. D., & Janusz, S. (2010). Challenges in Transportation Sector in Ghana. Journal Of Konbin, 4(16), 106–115.

[2] (2013). The World in 2013; ICT Facts and Figures. Geneva : ITU.

[3] World Bank. (2013). Country Profiles. Retrieved from http://www.world bank.org/en/country

[4] Aamefule, E. (2013). Nigeria attains 41,000km of optic fibre infrastructure. Retrieved 10 21, 2011, from Punch News: www.punchng.com/ business/nigeria-attains-41,000km-of-optic-fibre-infrastructure-minister

[5] Adeyemo, A. B. (2011). E-government implementation in Nigeria: An assessment of Nigeria's global e-gov ranking. Journal of Internet and Information System, 11–19.

[6] Adjin, D. M., & Tadayoni, R. (2011). ITS, As A Tool for Citizens In Developing. 8th EU ITS Conference Proceedings, Lyon - France.

[7] AfriCOG. (2010). Deliberate Loopholes- transparency lessons from the Privatisation of Telkom and Safaricom. Nairobi: Africa Centre for Open Governance (AfriCOG). Retrieved from Africa Centre for Open Governance (AfriCOG): http://www.africog.org/reports/Deliberate_loopholes.pdf

[8] Airzone One Ltd. (2011). Equatorial africa deposition network (EADN) measurement system descriptions and operations manual field operations (part A). (Operations Manual). Nairobi, Kenya:.

[9] Ajao, O. (2004). Interconnect Clog Ghana's Telecom Terrain. Retrieved from Techbaron: http://techbaron.com/interconnect-clog-ghanas-telecom-terrain

[10] Aker, J. C. (2011). "Dial "A" for agriculture: a review of information and communication technologies for agricultural extension in developing countries. Agricultural Economics, 42(6), 631–647.

[11] Aker, J., & Mbiti, I. (2010). Mobile phones and economic development in Africa. Center for Global Development Working Paper 211.

[12] Akwaja, C. (2013). Telecom Investment inflows hit 32 Billion. Lagos: Leadership Newspaper. Retrieved from http://leadership.ng/news/170713/telecoms-investment-inflows-hit-32bn

[13] Alhassan, A. (2007). Broken promises in Ghana's telecom sector. Retrieved from World Association for Christian Communication: http://www.waccglobal.org/en/20073-media-and-terror/459-Broken-promises-in-Ghanas-telecom-sector.html

[14] Alhomod, S. M., & Shafi, M. M. (2013). Mobile phone and e-government: the future for dissemination of public services. Universal Access in the Information Society, 1–6.

[15] All Africa. (2013). Kenya: Consortium Model Best for Kenya's LTE Deployment. Retrieved from All Africa: http://allafrica.com/stories/201305270024.html

[16] Ally, M. (2009). Mobile Learning: Transforming the Delivery of Education and Training. Athabasca University Press.

[17] Andreae, M. (1991). Biomass burning- its history, use, and distribution and its impact on environmental quality and global climate. Global Biomass Burning- Atmospheric, Climatic, and Biospheric Implications (A 92–37626 15–42). Cambridge, MA, MIT Press, 1991.

[18] Arnold, O., Richter, F., Fettwei, G., & Blume, O. (2010). Power Consumption Modeling of Different Base Station Types in Heterogenous Cellular Networks. Future Network and Mobile Summit. Stuutgard.

[19] ART. (2013). Telecommunication Regulatory Board of cameroon. Retrieved from http://www.art.cm/

[20] Ashok, S. (2007). "Optimizing Model for Community Based- Hybrid Energy System".

[21] Atkinson, R., Castro, D., & Ezell, S. J. (2009). The Digital Road to Recovery: A Stimulus Plan to Create Jobs, Boost Productivity and Revitalize America. The Information Technology and Innovation Foundation.

[22] Awotokun, K. (2005). Local Government Administration Under 1999 Constitution in Nigeria. Journal of Soc. Sci, 129–134. Retrieved 10 12, 2011, from http://www.krepublishers.com/02-Journals/JSS/JSS-10-0-000–000-2005-Web/JSS-10-2-077-147-2005-Abst-PDF/JSS-10–2-129-134-2005-188-Awotokun-K/JSS-10-2-129-134-2005-188-Awotokun-K.pdf.

[23] Babatunde , I. G., & Etal. (2012). Analytic study of the level of incursion of ICT in the Nigerian local governments. IJAR-CSIT, 1–24.

[24] Bank of American Merrill Lynch. (2012 Q2). Global wireless matrix.

[25] Bärnighausen, T., Chaiyachati, K., Chimbindi, N., Peoples, A., Haberer, J., & Newell, M. L. (2001). Interventions to increase antiretroviral adherence in sus-Saharan Africa: a SYSTEMATIC REVIEW OF EVALUATION STUDIES. Lancet Infectious Diseases 11, 942–51.
[26] Bartlett, S. (2011). Esoko and MTN put Farmers First in Ghana. Accra: Esoko Gh Ltd.
[27] Bartlett, S. (2011). Esoko Launches in Ghana. Ghana: Esoko Gh Ltd.
[28] Bartlett, S. (2011). Ground-Breaking Study Confirms famers Using Esoko Receives more for their Crops. Mali: Esoko.
[29] Batchelor, S., & Norrish, P. (2003). Sustainable Information Communication Technology (ICT). Gamos Ltd.
[30] Batchelor, S., Norrish, P., Scott, N., & Webb, M. (2003). Sustainable ICT Case Histories. Gamos Ltd.
[31] Beardsley, S., Enriquez, L., Bonin, S., Sandoval, S., & Brun, N. (2010). Fostering the Economic and Social benefits of ICT. Geneva: World Economic Forum.
[32] Benefo, K. D. (2004). The Mass Media and HIV/AIDS Prevention in Ghana. Journal of Health and Population on Developing Countries. Retrieved from http://www.jhpdc.unc.edu/
[33] Bhatti, B. (2012). Cyber Security and Privacy in the Age of Social Networks. In J. Zubairi, & A. Mahboob, Cyber Security Standards, Practices and Industrial Applications: Systems and Methodologies (pp. 57–47).
[34] Bhavnani, A., Chiu, R. W., Janakiram, S., Silarszky, P., & Bhatia, D. (2008). The role of mobile phones in sustainable rural poverty reduction.
[35] Biztech Africa. (2013). NCC chief assures of bill to protect telecoms infrastructure. Biztech Africa. Retrieved 10 21, 2013, from Biztech Africa: http://www.biztechafrica.com/article/ncc-chief-assures-bill-protect-telecoms-infrastruc/7021/?section=telecoms#.UmT9Y1Ckr Xo
[36] Blycroft Limited. (2012). Africa Mobile factbook.
[37] Boafo, O. (2009, 12 1). Esoko Ghana launches first agricultural commodity index. Retrieved 11 12, 2013, from http://www.modernghana.com/new s2/252085/1/esoko-ghana-launches-first-agricultural-commodity-.html: http://www.modernghana.com
[38] Boateng, K. O. (2007). ICT for sustainable development: what it is not,. Paper presented at the EuroAfrica-ICT Group Meeting, Brussels, Belgium.

[39] BOCRA. (2013, October 25). History Of BOCRA. Retrieved from http://www.bta.org.bw/history-bocra

[40] Bootsma, H. A., & Hecky, R. E. (2003). A comparative introduction to the biology and limnology of the African great lakes. Journal of Great Lakes Research, 29(4), 3–18.

[41] Bootsma, H. A. (C 2003). African great lakes. Ann Arbor, Mich.: International Association for Great Lakes Research. Bowen,.

[42] Bor, J. (2007). The Political Economy of AIDS Leadership in Developing Countries: An Exploratory Analysis. Social Science and Medicine, 64(8), 1585–599.

[43] Brand, P., & Schwittay, A. (2006). The missing piece: Human-driven design and research in ICT and development. Paper presented at the Information and Communication Technologies and Development, 2006. ICTD'06. Conference on, 2–10.

[44] Bresnahan, T., & Trajtenberg, M. (1992). General Purpose Technology: "Engine of Growth?". Massachusetts: National Bureau of Economic Research.

[45] BTC. (2013). Botswana Telecom Corporation. Retrieved from http://www.btc.bw/index.php?page=about_us/about

[46] Budde a. (2013, October 25). Botswana - Telecoms, Mobile and Broadband. Retrieved from Budde.com: http://www.budde.com.au

[47] Budde b. (2013). Cameroon - Telecoms, Mobile, Broadband and Forecasts. Retrieved from http://www.budde.com.au/Research/Cameroon-Telecoms-Mobile-Broadband-and-Forecasts.html

[48] Budde c. (2013). Cote d Ivoire (Ivory Coast) - Telecoms, Mobile, Broadband and Forecasts. Retrieved from Budde.com: http://www.budde.com.au/Research/Cote-d-Ivoire-Ivory-Coast-Telecoms-Mobile-Broadband-and-Forecasts.html

[49] Bwathondi, P. O., Ogutu-Ohwayo, R., & Ogari, J. (2001). Lake Victoria Fisheries management plan. In I. G. Cowx, & K. Crean. LVFRP/TECH/01/16, Technical Document No. 16.

[50] Calandro, E., Gillwald, A., Deen-Swarray, M., Stork, C., & Esselaar, S. (2012, December). Mobile Usage at the Base of the Pyramid in South Africa. An infoDev Publication prepared by Research ICT Africa and Intelecon.

[51] Caspary, G., & O'Connor, D. (2003). Providing Low-Cost Information Technology Access to Rural Communities in Developing Countries: What work? What pay? OECD.

[52] CCK. (2013). Communication Commission of Kenya. Retrieved from Communication Commission of Kenya: http://www.cck.go.ke/about/what_we_do.html

[53] Chaiyachati, K. H., Loveday, M., Lorenz, S., Lesh, S., Larkan, L. M., & Haberer, J. (n.d.). A Pilot Study of an m-Health Application for Healthcare Workers: Poor Uptake Despite High Reported Acceptability at a Rural South African Community-Based MDR-TB Treatment Program. PLoS ONE, 8(5).

[54] Chan, T. W., Rochelle, J., Sherry, H., Kinshuk, Sharples, M., Brown, T., & Norris, C. (2006). One-to-one technology-enhanced learning: An opportunity for global research collaboration. Research and Practice in Technology Enhanced Learning, 1(01). Retrieved from http://www.worldscientific.com/doi/abs/10.1142/S1793206806000032

[55] Chang, L. W., Kagaayi, J., Arem, H., Nakigozi, G., Ssempijja, V., Serwadda, D., & Reynolds, S. J. (2011). Impact of a m-Health Intervention for Peer Health Workers on AIDS Care in Rural Uganda: A Mixed Methods Evaluation of a Cluster-Randomized Trial. AIDS Behaviour, 15, 1776–1784.

[56] Chetty, M., Sundaresan, S., Muckaden, S., Feamster, N., & Calandro, E. (2013). Measuring Broadband Performance in South Africa. Paper submitted for ICTD Cape Town (under revision).

[57] Chiang, H. S., & Tsaur, W. J. (2011). Identifying Smartphone Malware Using Data Mining Technology. Proceedings of 20th International Conference on Computer Communications and Networks (ICCCN), (pp. 1–6).

[58] Chib, A., Wilkins, H., & Hoefman, B. (2013). Vulnerabilities in m-Health implementation: a Ugandan HIV/AIDS SMS campaign. Global Health Promotion.

[59] Clayton, J. (C2001). McGraw-hill illustrated telecom dictionary / (3rd ed.). New York: McGraw-Hill.

[60] Connected Urban Development. (2013). Green ICT infrastructure. Connected Urban Development.

[61] Correll, D. L. (1998). The role of phosphorus in the eutrophication of receiving waters: A review. Journal of Environmental Quality, 27(2), 261–266.

[62] Coutler, G. W. (1970). Population changes within a group of fish species in Lake Tanganyika following their exploitation. Journal of FishBiology, 2, 329–353.

[63] Crentil, P. (2013 a). From Personal to Public Use: Mobile Telephony as Potential Mass Educational Media in HIV/AIDS Strategies in Ghana. Suomen Antropologi: The Journal of the Finnish Anthropological Society, 38, 83–103.
[64] Crentsil , P. (2013 b.). Changing Healthcare Communication in Ghana. In Devy, Ganesh, G. Davis, & K. K. Charkravarty, Knowing Differently: The cognitive challenge of the indigenous (pp. 106–124). New Delhi: Routledge.
[65] Crentsil, P. (2001). Informal Communication and Health-seeking Behaviours in Ghana. (Master's thesis, University of Helsinki, Finland.
[66] Crentsil, P. (2007). Death, Ancestors, HIV/AIDS among the Akan of Ghana. Ph. D. dissertation, University of Helsinki, Research Series in Anthropology 10.
[67] Crentsil, P. (n.d.). From Personal to Public Use: Mobile Telephony as Potential Mass Educational Media in HIV/AIDS Strategies in Ghana. Suomen Antropologi: The Journal of the Finnish Anthropological Society, 38(1), 83–103.
[68] Crentsil, P. (In Press). Mobile technology and HIV/AIDS in Ghana. Ghana Studies Special issue.
[69] Crisp, B., & Swerissen, H. (2002). Program, agency and effect of sustainability in health promotion. Health Promotion Journal of Australia, 13, 40–42.
[70] Dada, J. (2007). Nigeria ICT Initiatives: Focus on participation. Retrieved from GisWatch: http://www.giswatch.org/en/country-report/civil-society-participation/nigeria & http://www.giswatch.org/sites/default/files/GISW_Nigeria.pdf
[71] Dale, J. W. (2001). Information Technology and the U.S. Economy. Harvard: Department of Economics, Harvard University.
[72] Danfulani, J. (2013). E-governance: A weapon for the fight against corruption in Nigeria. Retrieved from Sahara Reporters: http://saharareporters.com/article/e-governance-weapon-fight-against-corruption-nigeria-john-danfulani
[73] David-West, O. (2010). Esoko Networks,. UNDP.
[74] De Tolly, K., & Alexander, H. (2009). Innovative Use of CellphoneTechnology for HIV/AIDS Behaviour Change Communications: 3 pilot projects. Paper presented at the W3C conference on Africa Perspectives on the Role of Mobile Technologies in Fostering Social and Economic Development, 1–2 April. Retrieved from http://www.w3.org/2008/10/MW4D_WS/papers/kdetolly.pdf

[75] Delmas, R., Loudjani, P., Podaire, A., & Menaut, J. (1991). Biomass burning in africa- an assessment of annually burned biomass.Global Biomass Burning- Atmospheric, Climatic, and Biospheric Implications (A 92–37626 15–42). Cambridge, MA, MIT Press, 1991.

[76] Deloitte and GSMA. (2012). Sub - Sahara Africa, Mobile Observatory 2012.

[77] Deloitte LLP. (2012). Sub- Saharan Africa Mobile Observatory. GSMA.

[78] Donner, J. (2008). Research Approaches to Mobile Use in the Developing World: A Review of the Literature. The Information Society, 24(3), 140–59.

[79] Drayton, R. S. (1984). Variations in the level of Lake Malawi. Hydrological Sciences/Journal des Sciences Hydrologiques, 29, 1–12.

[80] Duda, A. (2002). Restoring and protecting the African Great Lake Basin ecosystems – lessons from the North American Great Lakes and the GEF. In E. O. Odada, & D. O. Olago, The East African Great Lakes: Limnology, Palaeolimnology and Biodiversity. Advances in Global Change Research (pp. 537–556). Kluwer Academic Publishers.

[81] Ebeling, M. (2003). The New Dawn: black Agency in Cyberspace. Radical History Review (98), 96–108.

[82] Ebrahim, H. (2013). Twitter Unrestricted File Upload Vulnerability. Retrieved 12 17, 2013, from http://securityaffairs.co/wordpress/19259/hacking/twitter-unrestricted-file-upload-vulnerability.html

[83] Eichler , S. (2007). Performance evaluation of the IEEE 802.11p WAVE communication standard. 66th IEEE Vehicular Technology Conference,(Baltimore, MD, USA) (pp. 2199–2033). IEEE.

[84] Ekine, S. (2012). SMS uprising: Mobile phone activism in Africa. Fahamu/Pambazuka.

[85] Emejor, C. (2012). Why 32b National rural telephongy project is abandoned. Retrieved 10 17, 2013, from Daily Independent: http://dailyindependentnig.com/2012/11/why-n32b-national-rural-telephony-project-is-abandoned/

[86] Ericsson. (2007). Sustainable Energy use in mobile communication. Ericsson White Paper.

[87] Essegbey, G. O., & Frempong, G. K. (2011). Creating space for innovation–The case of mobile telephony in MSEs in Ghana. Technovation, 31(12), 679–688. Retrieved from http://www.sciencedirect.com/science/article/pii/S0166497211001222

[88] EU Commission. (2013). Digital single market. Retrieved from http://ec.europa.eu/digital-agenda/en/our-goals

[89] European Commission. (2006). Effect of ICT on Economic Growth. Brussels: European Commission.

[90] Fearnside, P. M. (1985). Agriculture in Amazonia. Key Environments: Amazonia. Oxford, Reino Unido: Pergamon Press,.

[91] Felt, A. P., Finifter, M., Chin, E., Hanna, S., & Wagner, D. (2011). A survey of mobile malware in the wild. Proceedings of the 1st ACM workshop on Security and Privacy in Smartphones and Mobile Devices (SPSM), 2011.

[92] Fettweis, G., & Zimmermann, E. (2008). ICT ENERGY CONSUMPTION - TRENDS AND CHALLENGES. The 11^{rmth} International Symposium on Wireless Personal Multimedia Communications. Dresden, Germany: Wireless Personal Multimedia Communications.

[93] Forneld, D. M., Delaunay, G., & Elixmann, D. (2008). The Impact of Broadband on Growth and Productivity. Dusseldorf: European Commission.

[94] Forster, V., & Briceno-Garmendia, C. (2011). Africa's ICT infrastructure: Building on the Mobile Revolution. Washington: The World BanK Publications.

[95] Foth, M., & Hearn, G. (2007). Networked Individualism of Urban Residents: Discovering the Communicative Ecology in Inner City Apartment Buildings. Information Communication and Society, 10(5), 749–72.

[96] Frempong, G. (2002). Telecommunication Reforms –Ghana's Experience. Berichte aus dem Weltwirtschaftlichen Colloquium der Universität Bremen (pp. 1–49). Universität Bremen.

[97] Frempong, G. K. (2007). Restructuring of telecom sector in Ghana: Experiences and policy implications,. Centre for Information Technologies Technical University of Denmark, Lyngby,.

[98] Frempong, G. K. (2011). Ghana ICT sector performance review 2009/2010. Research ICT Africa, Cape Town, South Africa, 30 Pages.

[99] Frempong, G. K. (2012). Understanding what is happening in ICT in Ghana: A supply- and demand-side analysis of the ICT sector. Research ICT Africa Policy Paper 4, Cape Town South.

[100] Frohberg, D. (2006). Mobile Learning is Coming of Age: What we have and what we still miss. Proceedings of DeLFI, 327–338. Retrieved from http://www.comp.leeds.ac.uk/umuas/reading-group/MLearn_Framework .pdf

[101] Fuentelsaz, L., Maicas, J. P., & Polo, Y. (2008). The evolution of mobile communications in Europe: The transition from the second to the third generation. Telecommunications Policy, 32(6), 436–449.

[102] Gandhi, A. D., & Newbury, M. E. (2011). Evaluatinon of the energy efficiency metrics for wireless networks. Bell Labs Technical Journal, 207–215.

[103] Gartner. (2007). Gartner Estimates ICT Industry Account for 2 % of Global CO2 emission. Standford: Gartner.

[104] Gasmi, F., Maingard, A., Noumba, p., & Recuero Virto, L. (2011). Impact of privatization in telecommunications - A worldwide comparative analysis. Competition and Regulation in Network Industries.

[105] Ghana Population Census. (2010). Accra: Ghana Population Census.

[106] Ghana Statistical Service. (2012). Summary Report of Final Results. Accra: Ghana Statistical Service.

[107] Ghana Web. (1994). Transportation Problems. Retrieved from Ghana Web: http://www.ghanaweb.com/GhanaHomePage/transport/

[108] Ghana Web. (2005). Telephones and Communication. Retrieved from Ghana Web: http://www.ghanaweb.com/GhanaHomePage/communicati on/

[109] Gildert, P. (2006). Power System Efficiency in Wireless Communication. APEC 2006. Emerson Electric Co.

[110] Gillwald, A. (2004). Towards an african e-index: Household and individual ICT access and usage across 10 African Countries. Link Centre Wits University, School of Public and Development Management.

[111] Gillwald, A., Moyo, M., & Stork, C. (2013). Understanding What is Happening in ICT in South Africa, Evidence for ICT Policy Action series. Research ICT Africa(Policy Paper no. 7, 2012). Retrieved from www.researchICTafrica.net

[112] GIPC. (2013). Ghana Invesment Promotion Centre. Retrieved from Ghana Invesment Promotion Centre: http://www.gipcghana.com

[113] GISPA. (2012). Cost of Internet goes down . . . capacity goes up 65 times. Accra: GISPA. Retrieved from http://www.gispa.org.gh/news/?p=525

[114] GNA. (2013, June 10). Esoko, GNA to launch commodity price index. Retrieved 11 10, 2013, from http://www.ghanabusinessnews.com/2013/0 6/10/esoko-gna-to-launch-commodity-price-index/: http://www.ghanab usinessnews.com

[115] Goke. (2013). Galaxy Backbone emerges UN public service award winner. Retrieved 10 21, 2013, from Technology Times: www.ventures-africa.com/2003/06/galaxy -backbone-emerges-un-public-service-award s-winner/

[116] Gonzalez, M. C., Hidalgo, C. A., & Barabasi, A. (2008). Understanding individual human mobility patterns. Nature, 453(7196), 779–782.

[117] GRIDCo. Ghana. (2010). Ghana Wholesale Power Reliability Assessment. Accra: GRIDCo. Ghana.
[118] GSMA. (2011). African mobile observatory 2011: Driving economic and social development through mobile services, London, UK.
[119] GSMA. (2013). Sub-Saharan Africa Mobile Economy 2013. Retrieved 11 11, 2013, from GSMA Report: http://www.gsmamobileeconomyafrica.com
[120] GTP. (2013). Ghana Telecoms Policy. Retrieved from NCA: http://nca.org.gh/downloads/Ghana_Telecom_Policy_2005.pdf
[121] Guchteneire, P., & Mlikota, K. (2007). ICTs for Good Governance – Experiences from Africa, Latin America and the Caribbean, UNESCO. Retrieved July 7, 2012, from UNESCO: http://portal.unesco.org/pv_obj_cache/pv_obj_id_45CC12CF70DADF00E4124258B5E22CF0FC020100/filename/IST+Africa+paperrev.pdf
[122] Guemede, T., & plauche, M. (2009). Initial fieldwork for LWAZI: A telephone-based spoken dialog system for rural South Africa. In Proceedings of the First Workshop on Language Technologies for African Languages (pp. 59–65). Association for Computational Linguistics.
[123] Gupta, A., Calder, M., Feamster, N., Chetty, M, Calandro, E., & Katz-Bassett, E. (2013). Peering at the Internet's Frontier: A First Look at ISP Interconnectivity in Africa. Paper submitted to the PAM conference 2014, under revision.
[124] Gyepi-Garbrah, K. (n.d.). The perception of stakeholders in education on computerized school selection and placement system in Gomoa East and West districts in the Central region. Retrieved from http://ir.ucc.edu.gh/dspace/handle/123456789/1288
[125] H, M. (1979). Environmental chemistry of the elements.
[126] Haberer, J. E., Kiwanuka, J., Nansera, D., Wilson, I. B., & Bangsberg, D. R. (2010). Challenges in using mobile Phones for Collection of Antiretroviral Therapy Adherence Data in a Resource-Limited Setting. AIDS Behaviour, 14, 1294–1301.
[127] Haggarty, L., Shirley, M. M., & Wallsten, S. (2002). Telecommunication Reform in Ghana. ITU.
[128] Haggarty, L., Shirley, M., & Wallsten, S. (2002). Telecom Reform in Ghana. Washington DC: World Bank.
[129] Han, C. (2012). South African Perspectives on Mobile Phones: Challenging the Optimistic Narrative of Mobiles for Development. International Journal of Communication, 6(1), 2057–81.

[130] Hanek, G., Coenen, E. J., & Kotilainen, P. (1993). Aerial Frame Survey of Lake Tanganyika Fisheries FAO/FINNIDA Research for the Management of the Fisheries on Lake Tanganyika. GCP/RAF/271/FIN-TD/09.

[131] Harding , C. (2011). How Africa's economy is benefiting from the ICT revolution,. Retrieved December 30, 2013, from http://www.howwemadeitinafrica.com/how-africa%E2%80%99s-economy-is-benefitting-from-the-ict-revolution/12857/

[132] Hardy, A. P. (1980). The role of the Telephone in economic development. Telecommunication Policy, 278–286.

[133] Heeks, R. (2008). ICT4D 2.0: The next phase of applying ICT for international development. Computer, 41(6), 26–33.

[134] Heeks, R. B. (1999). Software Strategies in Developing Countries. COMMUNICATIONS OF THE ACM, 42.

[135] Heerden, A., Tomlinson, M., & Swartz, L. (2012). Point of care in your pocket: a research agenda for the field of m-health. Bulletin of World Health Organisation, 90, 393–394.

[136] Hellström, J., & Tröften, P. E. (2010). The innovative use of mobile applications in East Africa. Swedish international development cooperation agency (Sida).

[137] Hemen, T. P. (2012, June). Esoko and the future of agriculture in West Africa. Retrieved November 12, 2013, from http://westafricainsight.org/articles/PDF/181: http://westafricainsight.org

[138] Hirose, T., & Hirofuni, M. (2012). "standalone Hybrid Wind-Solar Power Generation System Applying Dump Power Control without Dump load". Transaction on Industrial Electronics.

[139] Hoekstra, D., & Corbett, J. (1995). Sustainable Agricultural Growth for Highlands of East and Central Africa: Prospects to 2020. Paper presented at the Ecoregions of the Developing World: A Lens for Asessing Food, Agriculture and the Environment to the Year 2020, held at Washington DC, USA. Organised by the International Food Policy Research Institute.

[140] Horwitz, R. (1999). South African Telecommunications: History and Prospects. In E. Noam, Telecommunications in Africa (pp. 205–248). New York: Oxford University Press. Retrieved from http://www.vii.org/papers/horwitz2.htm

[141] http://www.esoko.com/about/clients.php#casestudies. (2012). Retrieved July 10, 2013, from Esoko Website: http://www.esoko.com

[142] http://www.esoko.com/about/index.php#farmers. (2012). Retrieved July 10, 2013, from Esoko Website: http://www.esoko.com

[143] http://www.esokonigeria.com/newsreader.asp?news=1. (2013). Retrieve d 11 10, 2012, from http://www.esokonigeri a.com: http://www.esokonige ria.com

[144] http://www.esokonigeria.com/newsreader.asp?news=2. (2013). Retrieve d 11 10, 2013, from http://www.esokonigeria.com: http://www.esokonige ria.com

[145] Huang, Y. M., Chiu, P. S., Liu, T. C., & Chen , T. S. (2011). The design and implementation of a meaningful learning-based evaluation method for ubiquitous learning. Computers & Education, 57(4), 2291–2302. Retrieved from http://www.sciencedirect.com/science/article/pii/S03601 31511001291

[146] IBM. (2011). Delivering Intelligent Transport Systems -Driving integration and innovation. Retrieved from IBM: http://www-935.ibm.com/services/us/igs/pdf/transport-systems-white-paper.pdf

[147] IE Market Research Corp. (2011, November). Retrieved 11 12, 2013, from http://www.researchandmarkets.com/research/abdc85/3q11_ghana _mobile: http://www.researchandmarkets.com

[148] IEA. (2012). International energy agency, energy outlook. Retrieved from http://www.iea.org/publications/worldenergyoutlook/resources/ene rgydevelopment/accesstoelectricity/

[149] IFC. (2013). Kenya: Telkom Kenya. Washington: IFC. Retrieved from http://www.ifc.org/wps/wcm/connect/b5306380498391cc8614d6336b9 3d75f/SuccessStories_Telkom.pdf?MOD=AJPERES

[150] Infonetics. (2013). Service provider capex, opex, revenue, and subscribers database. Retrieved from http://www.infonetics.com/pr/2013/2 Q13-Service-Provider-Database-Highlights.asp

[151] Informa Telecoms & Media. (2013). Informa telecoms & media's world cellular information service. Retrieved March 2013, from http://Www.informatandm.com/contact-us

[152] Informa Telecoms and Media. (2012). Africa telecoms outlook 2013: Seizing new revenue opportunities. Retrieved from https://commerce.info rmatm.com/reports/africa-telecoms-outlook-2013.html

[153] International Telecommunications Union. (2010). Definitions of World Telecommunication/ICT Indicators. Retrieved August 20, 2013, from http://www.itu.int/ITU/ict/material/TelecomICT_Indicators_ Definition_ March2010_for_web.pdf

[154] International Telecommunications Union. (2013). Percentage of individuals using the internet. Retrieved November 11, 2012,

from http://www.itu.int/en/ITU-D/Statistics/Documents/statistics/2013/Individuals_Internet_2000–2012.xls
[155] Internet Society (ISOC). (2013). Report on Workshop on Reducing Internet Latency, 2013.
[156] Internet Society. (n.d.). The internet in Africa: A snapshot. Retrieved December 30, 2013, from http://internet-africa.projects.visual.ly/en/ on 30 December, 2013
[157] Internet World Statistics. (2011). Internet Usage for Africa. Retrieved June 9, 2011, from Internet world statistics: http://www.internetworldstats.com/stats1.htm
[158] Internet World Statistics. (2012). Internet Usage in Africa. Retrieved September 20, 2013, from Internet World Statistics: http://www.internetworldstats.com/stats1.htm
[159] Isaac, S. (2012a). Mobile Learning for Teachers In Africa and the Middle East. United Nations Educational, Scientific and Cultural Organisation (UNESCO).
[160] Isaac, S. (2012b.). Turning on Mobile Learning In Africa and Middle East. United Nations Educational, Scientific and Cultural Organisation (UNESCO).
[161] IT News Africa. (2012). Top ten largest telecoms companies in Africa. Retrieved November 12, 2013, from http://www.itnewsafrica.com/2012/08/top-ten-largest-telecoms-companies-in-africa/
[162] IT News Africa. (2013, February). Retrieved 10 12, 2013, from http://www.itnewsafrica.com/2013/02/ghana-mobile-users-top-25-million/: http://www.itnewsafrica.com
[163] ITU. (2004). Africa's Booming Mobile Markets: Can the Growth Curve Continue? Retrieved from ITU Telecom: http://www.itu.int/AFRICA2004/media/mobile.html
[164] ITU. (2010). Internet Usage Statistics. Geneva: ITU.
[165] ITU. (2010). Monitoring The WSIS Targets:A Mid-term view. World Telecommunication/ICT Development Report, Geneva.
[166] ITU. (2011). ICT facts and figures. Retrieved December 30, 2013, from http://www.itu.int/ITU-D/ict/facts/2011/material/ICTFactsFigures2011.pdf
[167] ITU. (2012). Measuring the information Society. International Telecommunications Union.
[168] ITU. (2013). IT Facts and figures. Retrieved from http://www.itu.int/en/ITU-D/Statistics/Pages/stat/default.aspx

[169] ITU STATSHOT. (2011). www.itu.int/ITU-ict/publications. Retrieved May 20, 2013, from www.itu.int/ITU-ict/publications.

[170] Jafaru, M. Y. (2012). Traffic Woes Over Soon - As BRT Project Begins. Retrieved from Modern Ghana: http://www.modernghana.com/news/315082/1/traffic-woes-over-soon-as-brt-pproject-begins.html

[171] Jagun, A., & Heeks, R. (2007). Mobile phones and development: The future in new hands? Id21 Insights, 69.

[172] Jeon, C. M., Amekudzi, A. A., & Vanegas, J. (2006). Transportation System Sustainability Issues in High-,Middle-, and Low-Income Economies: Case Studies from Georgia, U.S., South Korea, Colombia, and Ghana. Journal Of Urban Planning and Development, 182–192.

[173] JMP. (2013). WHO / UNICEF joint monitoring programme (JMP) for water supply and sanitation data and estimates. Retrieved from http://www.wssinfo.org/data-estimates/table/

[174] John, H. (2010). ICT and the Environment in Developing Countries. Advances in Information and Communication Technology(No.76), 236–247.

[175] Johnson , T. (1996). Sedimentary processes and signals of past climatic change in the large lakes of the East African rift valley. The Limnology, Climatology and Paleoclimatology of the East African Lakes.Gordon and Breach Amsterdam, 367–412.

[176] Jonathan, K. G. (2007). Estimating Total Power Consumption by Servers in the U.S and the World.

[177] Junglas, I., Abraham, C., & Watson, R. T. (2008). Task-technology fit for mobile locatable information systems. Decision Support Systems.

[178] Kaara, G. (2013). State failed public on Telkom Kenya sale deal. Retrieved from The People: http://www.thepeople.co.ke/16259/state-failed-public-on-telkom-kenya-sale-deal/

[179] Kajumbula, R. (n.d.). The effectiveness of mobile short messaging service (SMS) technologies in the support of selected distance education students of Makerere University, Uganda. Retrieved from http://pcf4.dec.uwi.edu/viewpaper.php?id=98

[180] Kalichman, S., & Simbayi, L. (2004). Traditional beliefs about the cause of AIDS and AIDS-related stigma in South Africa. AIDS CARE, 16(5), 572–580.

[181] Kalra, A., Rajiv , C., & Khanna, R. (2010). National Conf. on Computational Instrumentation NCCI 2010 CSIO , (p. Role of Zigbee Technology in agriculture sector). Chandigarh, India,.

[182] Kane, S. (2002). Telecom Reform and Poverty Alleviation in Kenya. Sabinet. Retrieved from http://reference.sabinet.co.za/webx/access/electronic_journals/afjic/afjic_n3_a3.pdf

[183] Karen, R., & McIntosh, G. (2012). Use of mobile devices in extension and agricultural production-a case study. Precision Agriculture, 63(32).

[184] Keegan, D. (n.d.). The incorporation of mobile learning into mainstream education and training. Retrieved from http://www.mlearn.org/mlearn2005/CD/papers/keegan1.pdf

[185] Keeling, B. (2010). PRESTAT AND ESOKO PARTNERSHIP. Retrieved 11 10, 2013, from http://www.prestat.co.uk/shop/trading-fairly: http://www.prestat.co.uk

[186] Kim, Y., Kelly, T., & Raja, S. (2010). Building broadband: Strategies and policies for the developing world. The World Bank, Global Information and Communication Technologies (GICT) Department. The World Bank.

[187] Knoche, H., & Huang, J. (2012). Text is not the enemy-how illiterates use their mobile phones. Paper presented at the NUIs for New Worlds: New Interaction Forms and Interfaces for Mobile Applications in Developing Countries-CHI 2012 Workshop.

[188] Kolavalli, S., Robinson, E., Diao, X., Alpuerto , V., Folledo, R., Slavova, M., Asante, F. (2012). Economic Transformation in Ghana; Where Will the Path Lead? International Food Policy Research Institute.

[189] Koutitas, G. (2010). Review of Energy Efficiency in Telecommunication networks. Telfor.

[190] Kozma, R. (2005). National Policies that Connect ICT- based Education reform to economic and social development. Human Technology, 1, 117–156.

[191] Kukulsa-Hulme, A., Traxler, J., & Pettit, J. (2007). Designed and user-generated activity in the mobile age (Vol. 2). Retrieved from http://oro.open.ac.uk/8080/

[192] Kumarm, P., & et al. (2009). ITS for Developing Countries , World Review of Intermodal. World Review of Intermodal Transportation Research, 2 (2–3), 201–207.

[193] Kuppusamy, M., Raman, M., & Lee, G. (2009). Whose ICTs Investment matters to Economic Growth: Private or Public? The Malaysian perspective. Electronic Journal of Information Systems in Developing countries, 1–19.

[194] Lazauskaite, V. (2008). International gateway liberalization. WSIS Facilitation Meeting on Action Line C6 ITU Headquarters,. Geneva: ITU.

[195] LDPOST. (n.d.). The history of prepaid phone cards. Retrieved from http://www.ldpost.com/telecom-articles/The-History-of-Prepaid-Phone-Cards.html

[196] Leon, N., Schneider, H., & Daviaud, E. (2012). Applying a framework for assessing the health systems challenges to scaling up m-Health in South Africa. BMC.

[197] Lewis, P. J., & Townson, C. W. (2004). Using Actor Network Theory ideas. In Iinformation Systems Research: a case study of action research.

[198] Longe, P. (2013). Nigeria - telecoms, mobile, broadband and forecasts. Retrieved from BuddeComm: http://www.budde.com.au/Research/Nigeria-Telecoms-Mobile-Broadband-and-Forecasts.html#overview

[199] Luc, D. (2005). Governance and Customs Operations: The Role of ICT. Regional Conference on Investment at Competitiveness.

[200] Luk, R., Zaharia, M., Ho, M., Levine B., & Aoki, P. M. (2009). ICTD for Healthcare in Ghana: Two Parallel Case Studies. ICTD '09 Proceedings of the 3rd International Conference on Information and Communication Technologies and Development. N.J.: IEEE Press: IEEE Press Picataway,.

[201] Luk, R., Zaharia, M., Ho, M., Levine, B., & Aoki, P. M. (2009). ICTD for Healthcare in Ghana: Two Parallel Case Studies. ICTD '09 Proceedings of the 3rd International Conference on Information and Communication Technologies and Development. IEEE Press Picataway, N.J IEEE Press.

[202] LusakaTimes. (2010). ZAMTEL privatization, an African first. Retrieved from Lusaka Times: http://www.lusakatimes.com/2010/08/17/zamtel-privatization-african/

[203] Majaliwa, M. J., Magunda, M. K., & Tenywa, M. (2003). Effect of contour bunds on soil erosion from major agricultural land-use systems in selected micro-catchments of Lake Victoria Basin. Paper presented at 18 SSSEA Conference, December 3rd–9.

[204] Market, K. (2005). Low-key m-learning: a realistic introduction of m-learning to developing countries Seeing, Understanding, Learning in the Mobile Age, Budapest, Hungar. Retrieved from http://www.fil.hu/mobil/2005/Masters_final.pdf

[205] Markus, M., & Robey, D. (1988). Information technology and organizational change: causal structure in theory and research. Management Science, 34, 583–598.

[206] Matthias, N. M. (2011). Assessing the Communicative Ecology of Male Refugees in Namibia: A Study to Guide Health Communication Interventions on Multiple and Concurrent Sexual Partnerships. Master's thesis University of Ohio.

[207] McAfee. (2012). 2012 Threat. Retrieved from McAfee: http://www.mcafee.com/us/resources/reports/rp-threat-predictions-2012.pdf

[208] McKinsey & Company. (2006). Wireless Unbound: The Surprising Economic Value and Untapped Potential of the mobile phone. McKinsey & Company.

[209] McQueen , S., Konopka, S., Palmer, N., Morgan, G., Bitrus, S., & Okoko, L. (2012). m-Health Compendium Edition One. In African Strategies for Health Project. Arlington VA.

[210] Meadows, K., & Zwick, K. (2000). The SESS recommendations to the Strategic Action Programme. Report No. IV, Final Pollution Control and Other Measures to Protect Biodiversity in Lake Tanganyika UNDP/GEF/RAF/92/G32.

[211] Melody, W. (2009). Markets and policies in new knowledge economies. In C. Avgerou, R. Mansell, D. Quah, & R. Silverstone, The Oxford Handbook of Information and Communication Technologies. Oxford Handbook.

[212] Melody, W. H. (1997). Policy Objectives and Models of Regulation. In W. H. Melody, Telecom reform: principles, policies and regulatory practices. Lyngby: Den Private Ingeniørfond, Technical University of Denmark.

[213] Minges, M. (1998). African telecoms: Private sector to the rescue? African Telecom Indicators. Retrieved from http://www.itu.int/ITU-D/ict/papers/bmi/bmi98.pdf

[214] Mitchell, K. J., Bull, S., Kiwankuka, J., & Ybarra, M. L. (2011). Cell phone usage among adolescents in Uganda:acceptability for relaying health information. Health Education Research, 26(5), 770–781.

[215] Mittal, S., Gandhi, S., & Tripath, G. (2010). Socio-economic impact of mobile phones on Indian agriculture. New Delhi,: : Indian Council for Research on International Economic Relations.

[216] Mobile News. (2008). Nigeria reaps from telecom licencing. Retrieved from Biz Community : http://www.bizcommunity.com/Article/157/78/23112.html

[217] MOFA. (2010). Agricultural Extension Approaches Being Implemented in Ghana. Accra: Directorate of Agricultural Extension Services; Ministry of Agriculture.

[218] Motorola. (n.d.). Nigerian GSM Operator M-tel Makes the Impossible Link Ayangba, Nigeria. Motorola.

[219] Müller-Kuckelberg , K. (2012). Climate Change And Its Impact On The Livelihood Of Farmers And Agricultural Workers In Ghana. Accra: Friedrich Ebert Stiftung.

[220] Mungai, W. (n.d.). Using ICTs for poverty reduction and environmental protection in kenya. M-Vironment" Approach.". In M. Andejelkovic, A Developing Connection: Bridging the Policy Gap between the Information Society and Sustainable Development. Winnpeg, Canada: IISD.

[221] Mursu, A. (2002). Information Systems Development in Developing Countries: Risk Management and Sustainability Analysis in Nigeria Software Companies. University of Jyvaskyla.

[222] Muwonge, J., & Gomes, E. (2007). Analysis of the Acquisition Process of Uganda. Management of International Business & Economic Systems, 1(1), 108–129.

[223] Naik, V. R., Padaria, R., Chandrashekar, N., & Babu, N. N. (2012). Mobile—A Catalyst in the Transfer of Agriculture Technology. Proceedings of M4D 2012 28–29 February 2012, 28(29), p. 151. New Delhi, India.

[224] Nankani, G. (2011). The Challenges of Agriculture in Ghana, What is to be Done? London: International Growth Center.

[225] NCA. (2013, September 20). http://www.nca.org.gh/73/34/News.html?item=332. Retrieved December 1, 2012, from http://www.nca.org.gh: http://www.nca.org.gh

[226] NCA. (2013). National Communications Authority. Retrieved from National Communications Authority: www.ncc.org.gh

[227] NCA a. (2003). National Communications Authority. Retrieved from THE GHANA ICT FOR ACCELERATED DEVELOPMENT: http://www.nca.org.gh/downloads/Ghana_ICT4AD_Policy.pdf

[228] NCC. (2012). Accelerating broadband penetration through USPF support. NCC. Retrieved 10 21, 2013, from NCC: http://www.ncc.gov.ng/index.php?option=com_content&view=article&id=949%3Aguardian-1st-august-2012-accelerating-broadband-penetration-through-uspf-support-&catid=86%3Acat-mediapr-headlines&Itemid=107

[229] NCC. (2013). National Communications Commission. Retrieved from http://www.ncc.gov.ng

[230] News. (n.d.). Retrieved from http://ghana.worlded.org/News.html

[231] NIPC. (2013). Nigerian investment Promotion Coporation. Abuja: NIPC. Retrieved from http://www.nipc.gov.ng/investment.html

[232] Nkemachor, B., & Nnadozie, J. (2013). Nigeria, ten million mobile phones for farmers. Retrieved from Vanguard: http://www.vanguardngr.com/2013/02/nigeria-ten-million-mobile-phones-for-farmers/

[233] Ntiba, M. J. (2003). Capacity building for shared water ecosystems in the Lake Victoria region. Tokyo, Japan: In press, United Nations University Press.

[234] Ntonzi, J. P. (1997). AIDS morbidity and the role of the family inpatient care in Uganda. Health Transition Review, 7, 1–22.

[235] Odada, E. O. (2006 c). Global international waters assessment : Regional assessment 47, East African rift valley lakes. Kalmar, Sweden: University of Kalmar on behalf of United Nations Environment Programme.

[236] Odemwingie, E. (2013). Nigeria: Pentascope - Ex-NITEL Staff Slam El-Rufa'i. Retrieved from All Africa: http://allafrica.com/stories/201304040578.html

[237] OECD. (2002). OECD Information Technology Outlook: ICTs and the Information Economy. France, Paris.

[238] OECD. (2004). ICTs AND ECONOMIC GROWTH IN DEVELOPING COUNTRIES. DAC Network on Poverty Reduction.

[239] OECD. (2009). Sustainable Manufacturing and Eco-Innovation, Synthesis Report, Framework, Practice and Measurement.

[240] Ogbomo, O. M. (2009). Information and Communication Technology (ICT) in Local Givernment Administration: The Case of Oshimili North Local Government Area of Delta State. Library Philosophy and Practice (Ejournal), 286. Retrieved from http://digitalcommons.unl.edu/libphilprac/286 & http://digitalcommons.unl.edu/libphilprac/286/

[241] Ogunlade Davidson and Standford A. Mwakasonda. (2004). Electricity access for the Poor: a study of South Africa and Zimbabwe. Energy for Sustainable Development 8 no., no.8.

[242] Okwuke, E. (2012). Nigeria's ARPU to decline by 6.9% this year- BMIL Report. Lagos: Daily Independent Newspaper. Retrieved from http://dailyindependentnig.com/2012/08/nigerias-arpu-to-decline-by-6–9-this-year-bmil-report

[243] Olabode, O., & Akingbesote, A. (2007). Embedding information and communications Technologies (ICT) in Nigeria Local Government System. Pacific Journal of Science and Technology, 8(2), 385–391.

[244] O'Loughlin, J., Renaud, L., Richard, L., Gomez, L. S., & Paradis, G. (1998). Correlates of the Sustainability of Community-based Heart Health Promotion Interventions. Preventive Medicine, 27(1), 702–712.

[245] Olufemi, J. F. (2012). Electronic Governance: Myth or Opportunity for Nigerian Public Administration? International Journal of Academic Research in Business and Social Sciencec, 2(9), 122–140. Retrieved 10 21, 2013, from http://www.hrmars.com/admin/pics/1104.pdf

[246] Owusu, K., & Waylen, P. (2009). Weather. Trends in spatio-temporal variability in annual rainfall in Ghana (1951–2000), 65(4), 115–120.

[247] Owusu, K., & Waylen, P. R. (2012). Theoretical and Applied Climatology. The changing rainy season climatology of mid-Ghana, 1–12.

[248] Owusu, K., Waylen, P., & Qiu, Y. (2008). GeoJournal. Changing rainfall inputs in the Volta basin: implications for water sharing in Ghana, 71(4), 201–210.

[249] Pamadas, M., Nallapernunal, D. K., Mualidharan, V., & Ravikumar, P. (2010). A Deployable Architecture of ITS – A Developing Country Perspective. In Computational Intelligence and Computing Research (ICCIC) 2010 IEEE International Conference, (pp. 1–6).

[250] Phippard, T. M. (2012). The (M)Hhealth Connection: An Examination of the Promise of Mobile Phones for HIV/AIDS Intervention in sub-Saharan Africa. Master's thesis. University of Western Ontario, Canada.

[251] PIDG. (2013). Developing Telcoms in Kenya. Retrieved from Private Infrastructure Development Group: http://www.pidg.org/impact/case-studies/kenya-telecoms

[252] Pinch, T., & Bijker, W. (1992). Shaping Technology/ Building Society. In W. Bijker, & J. Law. Cambridge, MA: MIT Press.

[253] Pool, R., Kamali, A., & Withworth, J. A. (2006). Understanding sexual behaviour change in rural southwest Uganda: A multi-method study. AIDS CARE , 479–488.

[254] Poptech. (2012). Project Masiluleke 2012. Retrieved from http://poptech.org/project_m

[255] Power Systems Energy Consulting (PSEC)/Ghana Grid Company (GRIDCo). (2010). Ghana Wholesale Power Reliabiliy. Accra: GRIDCo.

[256] Pulse, U. G. (2012). Big data for development: Challenges and opportunities. Naciones Unidas, Nueva York, Mayo,.

[257] Radstake, M. (1997). Secrecy and Ambiguity: Home Care for People Living with HIV/AIDS in Ghana. Leiden: African Studies Centre. Research Report 59/2000.

[258] Rebecca Mayer. (2007). ITU - D Study Group 2 Question 10- GSM 2G Base Station Power Consumption. Geneva: ITU.

[259] Reddy, P. K., & Ankaiah, R. (2005). A framework of information technology-based agriculture information dissemination system to improve crop productivity. Current Science, 88(12), 1905–1913.

[260] Research ICT Africa. (2012). Household and Small Business Access & Usage Survey 2011. Research ICT Africa. Retrieved from http://www.researchictafrica.net/publications/Research_ICT_Africa_Policy_Briefs/2012_Stork_-_ICT_Survey_Methodology.pdf.

[261] Research ICT Africa. (2013a). Benefits of reduced termination rate finally kick in with lower mobile prices, Policy Brief no. 1. Research ICT Africa. Retrieved from http://www.researchictafrica.net/docs/SA_Policy_Brief_2013_No_1_FINAL-WEB%20VERSION.pdf

[262] Research ICT Africa. (2013b). How do mobile and fixed broadband stack up in SA? RIA Policy Brief South Africa No. 2, July 2013. Research ICT Africa. Retrieved from http://www.researchictafrica.net/docs/SA_Policy_Brief_2013_No_2_%20July2013%20final%20webversion.pdf

[263] Ribbink, A. J. (n.d.). Lake Malawi/Niassa/Nyasa Ecoregion: Biophysical Reconnaissance. WWF Southern African Regional Programme Office (WWF SARPO). Harare, Zimbabwe.

[264] Roeth, H., & Wokeck, L. (2011). ICT and Climate Change Mitigation in Emerging Economies.

[265] Roller, L. H., & Waverman, L. (2001). Telecommunications infrastructure and economic development: a simultaneous approach. American Economic Review, 91(4), 909–923.

[266] Roschelle, J., & Pea, R. (n.d.). A walk on the WILD side: How wireless handhelds may change CSCL. In Proceedings of the Conference on Computer Support for Collaborative Learning Foundations for a CSCL Community. Retrieved from http://dl.acm.org/citation.cfm?id=1658624

[267] Safaricom. (2013). Our history our heritage. Retrieved from Safaricom: http://www.safaricom.co.ke/about-us/about-safaricom/our-history-heritage

[268] Sahara Reporter. (2013). Multi-million Naira rural IT centers abandoned. Retrieved from Sahara Reporters: www.saharareporters.com/news-page/multi-million-naira-rural-it-centres-abandoned

[269] Samadashvili , L. (2011). E-governance in Local Government, Civil Registry Agency. Retrieved 10 19, 2012, from http://www.slideshare.net/E-Gov_Center_Moldova/e-gov-and-local-gov-levan-samadashvili-eng-as-of-sept-20–9590444

[270] Sameer Group. (2013). Sameer in Telecommunications. Retrieved from Sameer Group: http://www.sameer-group.com/index.php?option=com_content&view=article&id=28&Itemid=53

[271] Sanganagouda, J. (2011). Ussd: A communication technology to potentially oust sms dependency.

[272] Sanner, T. A., Roland, L. K., & Braa, K. (2012). From pilot to scale: Towards an m-Health typology for low-resource contexts. Health Policy and Technology, 155–164.

[273] Santosh , A. S. (2012). Broadband spreading across Africa. Retrieved December 29, 2013, from http://www.other-news.info/2012/02/broadband-spreading-across-africa/ on 29 December 2013.

[274] Sengupta, D. V., & Bansal, D. A. (2012). AN EMPIRICAL STUDY ON USE OF ICT IN FRONT OFFICE OPERATION AS A TOOL TO IMPROVE SERVICE QUALITY AND INCREASED REN=VENUE GENERATION. A Journey of Economic and Management.

[275] Senyo, K. (2013, August 27). UMOVE And Stay Healthy, Ghana's Innovative Mobile App. Retrieved 10 10, 2013, from http://awakeafrica.org/archives/2833#sthash.DRJYYf8l.dpbs: http://awakeafrica.org

[276] Siedner, M. J., Lankowski, A., Musingi, D., Jackson, J., & Haberer, J. (2012). Optimizing Network Connectivity for Mobile Technologies in sub-Saharan Africa. PLoS, 7(9).

[277] Singh, A. (2006). Information and Communication Technologies (ICT) and Sustainable Development Alternatives. Retrieved 12 12, 2012, from www.devalt.org

[278] Singh, A. (2006). Information and Communication Technologies (ICT) and Sustainable Development Alternative. Retrieved from Devalt: www.devalt.org

[279] Sophos. (2012). Security Threat Report 2012. Retrieved November 11, 2013, from Sophos: http://www.securelist.com/en/analysis?pubid=204792051

[280] Spigel, R. H., & Coulter, G. W. (1996). Comparison of hydrology and physical limnology of the East African Great Lakes TanganyikaMalawi, Victoria, Kivu and Turkana (with reference to some North American Great Lakes). In T. C. Johnson, & E. O. Odada, The Limnology, Climatology and Paleoclimatology of the East African Lakes (pp. 103–139). Toronto: Gordon and Breach.

[281] Stork, C., Calandro, E., & Gamage, R. (2013). The Future of Broadband in Africa,. Research ICT Africa. Retrieved from www.resaerchICTafrica.net

[282] Stork, C., Calandro, E., & Gillwald, A. (2013). Internet going Mobile. Info, 15(5).

[283] Strand Consult. (2012). Next generation prepaid. Retrieved from http://www.strandconsult.dk/sw4152.asp

[284] Subair, G. (2013). 80% Nigerians set to enjoy rural telephony. Retrieved 10 17, 2013, from Nigerian Tribune: http://tribune.com.ng/news2013/index.php/en/component/k2/item/6701–80-nigerians-set-to-enjoy-rural-telephony

[285] Subervie, J. (2011). Evaluation of the Impact of A Ghanaian-Based MIS on first few users using Quasi-Experimental Design. Workshop on African Marketing Information Systems. Bamako: French National Institute for Agricultural Research (INRA).

[286] Sumbwanyambe, M., & Nel, A. L. (2011). Liberalization, Regulation and Privatization (LRP). 15th International Conference on Intelligence in Next Generation Networks (pp. 202–206). Berlin: IEEE Xplore.

[287] Swerissen, H., & Crisp, B. R. (2004). The Sustainability of Health Promotion Interventions for Different Levels of Social Organisation. Health Promotion International, 1, 123–130.

[288] Tamrat, T., & Kachnowski, S. (2001). Special delivery: An analysis of m-Health in maternal and newborn health programs and their outcomes around the world. Maternal Child Health Journal. Retrieved from http://www.advancefamilyplanning.org.

[289] Telkom Kenya. (2013). History of Telkom Kenya. Retrieved from Telkom Kenya: http://www.telkom.co.ke/index.php?option=com_content&view=article&id=60&Itemid=1

[290] Tenhunen, S. (2013). Introduction: Mobile Technology,gender and development. Suomen Antropologi: Journal of the Finnish Anthropological Society, 38(1), 4–11.

[291] Thakur, V. (2005). Implementing e-governance in Developing Countries-Key challenges. Paper presented at the E-government Summit ,Conflux 2005,. Delhi. Retrieved June 27th, 2012, from www.conflux.csdms.in

[292] The Boston Consulting Group. (2008). Smart 2020: Enabling the low carbon economy in the information age. U.S. Report Addendum.

[293] Thirumurthy, H., & Lester, R. T. (2012). M-health for health behaviour change in resource-limited settings: applications to HIV care and beyond. Bulletin of World Health Organisation, 5, 390–392.

[294] Thomas, D. (n.d.). Mideast operators join race for african telecoms. Retrieved from http://www.ft.com/intl/cms/s/0/ae54b460–37da-11e3–8668-00144feab7de.html?siteedition=intl

[295] Tomlinson, M., Rotheram-Borus, M. J., Swartz, L., & Tsai, A. C. (2013). Scaling up m-Health: where is the evidence. PLOS Med, 10(2).
[296] Traxler, J., & Kukulsa-Hulme, A. (2005). Evaluating mobile learning: Reflections on current practice. Retrieved from http://oro.open.ac.uk/12819/
[297] Traxler, J., & Leach, J. (2006). Innovative and sustainable mobile learning in Africa. In Wireless, Mobile and Ubiquitous Technology in Education, 2006. WMUTE'06 Fourth IEEE International Workshop (pp. 98–102). Fourth IEEE International Workshop. Retrieved from http://ieeexplore.ieee.org/xpls/abs_all.jsp?arnumber=4032531
[298] TRB. (2013). The Telecom Regulatory Board (ART). Retrieved from http://www.art.cm:81/index.php?option=com_content&task=view&id=18&Itemid=64
[299] U.S Energy Information. (2007). www.eia.doe.gov/. Retrieved July 14, 2011.
[300] UCC. (2013). Industrial Statistics. Retrieved from Uganda Communications Commission: ucc.co.ug
[301] UN. (2013). UN Economic report on Africa. Retrieved from http://www.uneca.org/publications/economic-report-africa-2013
[302] UN Statistics. (2009). http://mdgs.un.org/unsd/mdg/default.aspx. Retrieved June 1, 2010
[303] UNAIDS Regional Fact Sheet. (2012). Sub-Saharan Africa. Geneva: UNAIDS World Health Organization (WHO) 2005. WHO recommendations for clinical monitoring to support scale-up of HIV care, antiretroviral therapy and prevention in resource- constrained settings. Retrieved from http://www.who.int/hiv/pub/guidelines/clinical/monitoring.pdf
[304] UNCTAD. (2012). Mobile money for business development in the East African community. Retrieved from http://unctad.org/en/PublicationsLibrary/dtlstict2012d2_en.pdf
[305] UNDP. (2012). Mobile technologies and empowerment: Enhancing human development through participation and innovation. United Nations Development Programme.
[306] UNEP. (2011). Equatorial Africa deposition network (EADN) project description.
[307] United Nations. (2010). New and Emerging Technologies: Renewable Energy for Development. Geneva: United Nations.
[308] United Nations. (2013). Africa Renewal. Retrieved from http://www.un.org/africarenewal/topic/mobile-phones
[309] UNU-INWEH. (2011). Transboundary lake basin management: Laurentian and African great lakes. Hamilton, Ontario, Canada: UNU-INWEh.

[310] Vasumathi , S., Vaibhav , S., & Minaxi , G. (2012). Twitter games: how successful Spammers pick targets. In Proceedings of the 28th Annual Computer Security Applications Conference (ACSAC '12) (pp. 389–398). New York, NY: ACM.

[311] Vodafone. (2006). Vodafone Corporate Social Responsibility. Corporate Social Responsibility.

[312] Vodafone Ghana. (n.d.). About eLearning @ Vodafone. Retrieved from http://www.vodafone.com.gh/Internet-Cafes/Internet-cafe/eLearning-@-Vodafone-Ghana/About-eLearning-@-Vodafone.aspx

[313] WAECDIRECT ONLINE - RESULT CHECKER. (n.d.). Retrieved February 16, 2013, from http://www.waecdirect.org

[314] Wallsten, S. J. (2001). An Econometric Analysis of Telecom Competition, Privatization, And Regulation In Africa And Latin America. The Journal of Industrial Economics, XLIX, 1–19.

[315] Welsum, D. V. (2008). Broadband and the Economy. OECD.

[316] West, D. M. (2012). How Mobile Technology is Driving Global Entrepreneurship. Brookings Institute.

[317] WHO. (2011). m-Health: New horizons for health through mobile technologies,. Global Observatory for eHealth series - Volume 3, WHO, Geneva,. Retrieved from www.who.int/goe/publications/goe_m-Health_web.pdf?.

[318] Wicander, G. (2010). M4D Overview 1.0: The 2009 Introduction to mobile for Development.

[319] Williams, I. (2012). Infrastructure Development: Public Private Partnership Path for Developing Rural Telecommunications in Africa. Journal of Technology Management & Innovation, 7(2), 63–71.

[320] Wills, A., & Daniels, G. (2003). Nigeria Telecommunications Market-A Snap Shot View. Africa Analysis.

[321] Wilson, J. (2009). Challenges of Information and Communication Technologies (ICTs) In Promoting Democracy and Human Rights in Nigeria,. A paper presented at the 2009 Conference of Centre for Research and Documentation (Journalists and the Struggle for Democracy and Human Right) 7th–11th October 2009. Kano-Nigeria.

[322] World Bank .b. (2013). GNI per capita, Atlas method (current US$). Washington: World Bank. Retrieved October 11, 2013, from http://data.worldbank.org/indicator/NY.GNP.PCAP.CD?page=3

[323] World Bank. (1996). Kenya, Tanzania and Uganda: Lake Victoria Environmental Management Project. GEF Documentation. Report No. 15541 – ARF.

[324] World Bank. (2003). Malawi – Lake Malawi Ecosystem and Management Project. Africa Regional Office, Project ID MWPL-66196. Retrieved from http://www.cbnrm.net/pdf/worldbank_007_projrct_malawi_001_ecosystemgmt_pid.pdf

[325] World Bank. (2012). Maximizing mobile. Retrieved from http://go.worldbank.org/0J2CTQTYP0

[326] World Bank c. (2013). Literacy rate, adult total (% of people ages 15 and above). Washington: World Bank.

[327] World Bank. a. (2013). Urban population (% of total). Washington: World Bank. Retrieved October 11, 2013, from http://data.worldbank.org/indicator/SP.URB.TOTL.IN.ZS?page=3

[328] Yan, C., Zhang, S., Xu, S., & G. Y. Li. (2011). Fundamental trade-offs on green wireless networks. IEEE Magazine, 6.

[329] Young, S. (n.d.). African undersea cables- a history, 1999–2012. Retrieved December 27, 2013, from http://www.slideshare.net/ssong/african-undersea-cables-a-history

[330] Yu, P., Wu, M. X., Yu, H., & Xiao, G. Q. (2006). The Challenges for the Adoption of M-Health. Paper presented at the IEEE International Conference on Service Operations and Logistics, and Informatics, 2006 SOLI '06. IEEE. Retrieved from http://ieeexplore.ieee.org/xpls/abs_all.jsp?arnumber=4125574

[331] Zi Chu, S. G., Haining, W., & Sushil , J. (2010). Who is tweeting on Twitter: human? Bot, or cyborg? In Proceedings of the 26th Annual Computer Security Applications Conference (ACSAC '10). (pp. 20–30). NEW York: ACM.

[332] Zuckerman, E. (2009). Web 2.0 tools for development: Simple tools for smart people. Participatory Learning and Action, 59(1), 87–94.

Acronyms

3D	Third Dimension
2G	Second Generation
3G	Third Generation
4G	Fourth Generation
ADSL	Asynchronous Digital Subscriber Line
AIDS	Acquired Immune Deficiency Syndrome
ALGON	Association of Local Government of Nigeria
AMPS	Advanced Mobile Phone System
ANT	Actor Network Theory
API	AIDS Programme Effort Index
API	Application Programming Interface
APTS	Advanced Public Transport systems
ARPU	Average Revenue per User
ART	Antiretroviral Treatment
ARVs	Antiretroviral drugs
ATMS	Advanced Traffic Management Systems
ATIS	Advanced Traveller Information Systems
AVHS	Advanced Vehicle & Highway Systems
BWA	Broadband Wireless Access
CANI	Computers for All Nigerians Initiative
CBS	Community-Based health Services
CD	Compact Disc
CDMA	Code Division Multiple Access
CPE	Customer Premise Equipment
CTS	Conventional Transportation Sectors
CVO	Commercial Vehicle Operations
DoS	Denial of Service Attacks
DDoS	Distributed Denial of Service Attacks
DSL	Digital Subscriber Line
DVD	Digital Video Disc
EBITDA	Earnings Before Interest, Taxes, Depreciation and Amortization
EDGE	Enhanced Data rates for GSM Evolution

E Health	Electronic Health
ETACS (TAC)	Total Access Communication System
ETC	Electronic Toll Collection
EU	European Union
EVDO	Evolution Data Optimized
FECA	Front End Collision Avoidance
FHI	Family Health International
GB	Gigabytes
Gbps	Gigabytes Per Second
GDP	Gross Domestic Product
GPS	Global Positioning System
GSM	Global System Mobile
HIV	Human Immunodeficiency Virus
ICT	Information Communications and Technology
ICT4D	Information Communications Technology For Development
IP	Internet Protocol
IPPIS	Integrated Payroll and Personnel Information System
ISP	Internet Service Providers
IT	Information Technology
ITS	Intelligent Transport Systems
ITU	International Telecommunications Union
IVTT	Intelligent Vehicle Tracking Technology
IXP	Internet Exchange Points
LAN	Local Area Network
LTE	Long Term Evolution
M Health	Mobile Health
MARPs	Most at Risk Persons
MBPS	Mega Bytes Per Second
MDG	Millennium Development Goals
MHz	Mega Hertz
MMS	Mobile Message Service
MNO	Mobile Network Operator
MVNOs	Mobile Network Virtual Operators
NACP	National AIDS/STI Control Programme
NCA	National Communications Authority
NCC	National Communications Commission
NCPI	National Commitments and Policies Institution
NeGSt	National e-Governance Strategies
NFC	Near Field Communication

NITDA	National Information Technology Development Agency
NGO	Non-Governmental Organization
NRTP	National Rural Telephony Project
OS	Operating System
OTT	Over The Top
P&T	Post and Telecoms
PC	Personal Computer
PDA	Personal Digital Assistant
POTS	Plain Old Telephone Systems
PPP	Public Private Partnerships
PRTS	Public & Rural Transportation Systems
PSTN	Public Switch Telephone Networks
PTDF	Petroleum Technology Development Fund
RAN	Radio Access Network
RIA	Research ICT Africa
RITC	Rural Information Technology Centres Project
RTT	Round Trip Time
SHARPER	Strengthening HIV/AIDS Response Partnership with Evidence-Based Results
SIM	Subscriber Identification Module
SMS	Short Message Service
SPV	Special Purpose Vehicle
SSL	Secure Socket Layer
STI	Sexual Transmitted Infection
TIMS	Traffic Information Management Systems
TIS	Traveller Information Systems
UMTS	Universal Mobile Telecommunications System
UNAIDS	United Nations Programme on HIV/AIDS
UNDP	United Nations Development Programme
UNESCO	United Nations Educational, Scientific and Cultural Organization
URL	Uniform Resource Locator
USAID	U.S. Agency for International Development
USD	US Dollars
USSD	Unstructured Supplementary Service Data
USPF	Universal Service Provision Fund
Wi-Fi	Wireless Fidelity
WiMAX	Worldwide Interoperability for Microwave Access
WiN	Wire Nigeria
WHO	World Health Organization

Index

About the Editor

Prof. Knud Erik SKOUBY is Professor and Director of center for Communication, Media and Information Technologies, Aalborg University-Copenhagen. Has a career as a university teacher and within consultancy since 1972. The working areas have for the last 25 years been within the telecom area focusing on mobile/ wireless:*Techno-economic Analyses; Development of mobile/ wireless applications and services: Regulation of telecommunications*

Project manager and partner in a number of international, European and Danish research projects. Served on a number of public committees within telecom, IT and broadcasting; as a member of boards of professional societies; as a member of organizing boards, evaluation committees and as invited speaker on international conferences; published a number of Danish and international articles, books and conference proceedings in the areas of convergence, mobile/wireless development, telecommunications regulation, technology assessment; demand forecasting and political economy. Board member of the Danish Independent Research Council and the Danish Media Committee. Chair of WGA in Wireless World Research Forum; Special Advisor to GISFI (The Global ICT Standardization Forum for India).

Idongesit WILLIAMS is a PhD Fellow and a guest lecturer at Center for Communication Media and Information Technologies (CMI), Aalborg University. He holds a bachelor in Physics, a Master degree in Information Communications and Technology. His research has been specialized on the collaboration of Public and Private Institutions in the development of ICTs. He has published academic papers on Public Private Interplay in developing telecom network infrastructure for rural areas and presented papers at ICT conferences.

About the Authors

Daniel Michael Okwabi ADJIN holds a Master's Degree in Telecommunications Engineering. He is currently a Ph.D. Fellow in Telecommunications Engineering at Aalborg University Copenhagen, Denmark. He is a Telecommunications Engineer by profession and currently a Lecturer at Ghana Technology University College (GTUC), Accra, and a supervisor to many Undergraduate and Postgraduate Students at GTUC, and periodically, an adjunct Lecturer at Legon University of Ghana.

Dr. Nana Kofi ANNAN is a lecturer at Wisconsin International University College Ghana, Accra. He is attached to the Information Technology Department. He has a first degree in Computer Science and Management, a second degree in adult education and management all from Ghana, and PhD degree in Information and Communication Technology from Aalborg University Denmark. He has six years teaching experience as a university lecturer and over sixteen years of professional experience as a project manager. His research interest includes mobile technologies, mobile computing, mobile learning, educational technologies and communication technologies. His institutional address is Wisconsin International University College Ghana, P. O. Box LG 751, Legon – Accra.

Enrico CALANDRO is a research fellow at Research ICT Africa. Prior to joining Research ICT Africa, he worked as a technical advisor for the ICT program of the SADC Parliamentary Forum in Namibia, within the UN technical cooperation framework. He also worked for the European Commission, Information Society and Media DG as a trainee after completing his Master degree in Communications' sciences, University of Perugia, Italy. Currently, Enrico is a PhD candidate at the Graduate School of Business, University of Cape Town, program on Managing Infrastructure Reform and Regulation. He is a recipient of the Amy Mahan PhD scholarship award for the advancement of ICT policy and regulatory research in Africa, of the UNDESA fellowship for International cooperation and of the Emerald Literati Network Award for Excellence 2013.

Dr. Perpetual CRENTSIL is a postdoctoral researcher at the Social and Cultural Anthropology, University of Helsinki, Finland. Her areas of research interest are HIV/AIDS in Ghana (Africa), reproductive health, African medical systems, and kinship. Other research areas are migration and African Diaspora, focusing on remittances, gambling, gender and identities among African (and Asian) migrants in Finland. Her current research focuses on mobile technology, gender and health development in Africa. Her research project, 'Mobile technology and healthcare delivery services in Ghana' was a part of the larger Academy of Finland-funded research project, 'Mobile technology, gender and development in Africa, India and Bangladesh' (2010-2013).

Dr. Godfred FREMPONG is a Principal Research Scientist of the CSIR-Science and Technology Policy Research Institute (STEPRI). He holds a Ph.D. in Sociology from the University of Ghana and an M.Soc Sci. in Science Policy from the University of Lund (Sweden). He has also received post-qualification training from the University of Amsterdam, Wageningen University, The Centre for Information Technologies at the Technical University of Denmark and PREST of Manchester University (as a Commonwealth Professional Fellow in 2006). He has written extensively on ICT reform, regulation and development, and has rich experience in ICT policy research as well as a deep knowledge of telecommunications and ICT development in Ghana. Godfred Frempong also coordinates Research ICT Africa Network's research activities in Ghana.

Nuhu GAPSISO is a lecturer at the Department of Mass Communication, University of Maiduguri, Nigeria. His teaching and research interest is in area of Health Communication, Public Relations and Journalism. . His most recent research (yet to be published) is on Media Coverage of Human Rights Issues in Nigeria. He is a member of African Council for Communication Education. He is a member of Faculty of Social Sciences, Board of Examiners, University of Maiduguri, Nigeria. He has served as resource person and facilitator in several workshops in Nigeria. Email address: ndgapsiso@yahoo.com. Tel: +2348034315048, +2348054354369

Prof. Alison GILLWALD is Executive Director of Research ICT Africa, an ICT policy and regulatory think-tank based in Cape Town, South Africa, which hosts an Africa-wide research network. She is also adjunct professor at the Management of Infrastructure Reform and Regulation Program at the University Of Cape Town Graduate School Of Business. Prior to this she was Associate Professor at the Witwatersrand University's Graduate School of Public and Development Management, where she founded the Learning Information Networking and Knowledge (LINK) Centre in 1999. She served on the founding Council of the South African Telecommunications Regulatory Authority (SATRA) and the first Independent Broadcasting Authority prior to that. She has chaired the national Digital Broadcasting Advisory Task Team and served on the African Communications Ministers' Expert Panel. She has advised and been commissioned by regional bodies, governments, regulators and competitions commissions on the continent and multilateral agencies, including the African Development Bank, infoDev, the Commonwealth Telecommunication Organization, the International Telecommunications Union and ICANN. She is founding editor of the African Journal of Information and Communication and is published in the area of telecommunications and broadcasting policy and regulation, gender and political economy more generally.

Dr. Patrick Ohemeng GYAASE is Lecturer at the Faculty of Information, Communication Sciences and Technology (ICST), Catholic University College of Ghana, Fiapre, Sunyani since August, 2004. He also works as a Facilitator of Management Information Systems and E-commerce at Kwame Nkrumah University of Science and Technology (Centre for Distance Learning) He obtained his first Degree in Business Administration (Accounting Option), from the University of Ghana, Legon, A Masters Degree in Information Technology from UWE, Bristol in the UK and PhD in Information Technology at CMI, Department of Electronic Systems; Aalborg University, Denmark. His research interests are in e-Government, e-Business and information systems development and diffusion in the public sector. Email: pakw@es.aau.dk,pkog@cug.edu.gh

Prof. Anders HENTEN (MSO) is Professor at center for Communication, Media and Information technologies (CMI) at the Department of Electronic Systems at Aalborg University in Copenhagen. He is a graduate in communications and international development studies from Roskilde University in Denmark (1989) and holds a Ph.D. from the Technical University of Denmark

(1995). He has worked professionally in the area of communications economy and policy for more than 25 years. He has participated in numerous research projects financed e.g. by the European Community, the Nordic Council of Ministers, Danish Research Councils and Ministries, and in consultancies, financed by World Bank, UNCTAD, ITU, Danish Ministries, etc. He has published nationally and internationally – more than 250 academic publications in international journals, books, conference proceedings, etc.

Abdullahi ISAH holds a Master's degree in Information Technology, with specialization in Network security. He has been a lecturer at Hupoly Katsina state, Nigeria, for over twenty years. He is now pursuing a Ph.D. degree at CIMI – Department of electronic System, Aalborg University, Denmark. As a Ph.D. fellow, his main area of research is on the prevention of attacks on Wireless LAN, through the user being the weakest link in security controls. Abdullahi has been offering advices and consultancy services to individuals, students, and organizations within the African environment, on proper user security habits with network, devices and web services. His goal is to minimize targeted attacks through human vulnerabilities and to counteract malware and phishing threats. His educational/academic and research backgrounds, coupled with his keen interest and experiences in Social engineering, have made him an eagle eye in human vulnerabilities and attacks.

Prof. Gail KRANTZBERG is Professor and Director of the Arcelor Mittal Dofasco Centre for Engineering and Public Policy at Mc Master University. The Centre offers a Master's Degree to engineers and scientist training them to understand the application of science and technology to public policy. Gail completed her Ph.D. at the University of Toronto on contaminant cycling in freshwaters. She worked for the Ontario Ministry of Environment from 1988 to 2001, as Coordinator of Great Lakes Programs, and Senior Policy Advisor on Great Lakes. She is the past president of the International Association of Great Lakes Research. Dr. Krantzberg was the Director of the Great Lakes Regional Office of the International Joint Commission from 2001 to 2005. She is a board member of the Canadian Association for Water Quality, Georgian Bay Forever, and the Clean Water Foundation,. Gail has authored 5 books and more than 150 scientific and policy articles on issues pertaining to ecosystem quality and sustainability to promote environmental protection and social welfare.

Benjamin KWOFIE is lecturer and researcher at the department of Computer Science, School of Applied Science and Technology at the Koforidua Polytechnic, Ghana. His teaching and research interests are in the fields of Information Systems and organizations, technology and education, research and innovation in education, telecommunications, and policy and community development using ICT. He is currently pursuing a PhD in e-learning implementation in higher education institutions at the Aalborg University, Denmark.

Alejandro ISLAS LOPEZ is a Technology/Policy consultant. Alejandro completed his M. Eng. in Public Policy at Arcelor Mittal Dofasco Centre for Engineering and Public Policy at McMaster University and a B.S. equivalent in Telecommunications Engineering at the National Autonomous University of Mexico. From 2003 to 2012, he worked as a Software Engineer and a Systems Analyst developing Machine to Machine (M2M) solutions related to different topics such as electricity or water conservation. As a former United Nations University fellow in the Water without Borders Graduate Diploma, he became interested in the application of Information and Communications Technologies for Development.

Roslyn LAYTON, American, is a Ph.D. Fellow in internet economics at the Center for Communication, Media and Information Studies at Aalborg University in Copenhagen, Denmark and an employee of Strand Consult, an independent consultancy working with the mobile industry. Roslyn previously worked in the software industry in the US, India and Europe.

Dr. George Orleans OFORI-DWUMFOU is a Senior Lecturer at the Methodist University College Ghana, Accra where he heads the Department of Information Technology. Dr. Ofori-Dwumfuo has a first degree in Mathematics from Ghana, and a PhD. Degree in Computer Science from the United Kingdom. He has over twenty years experience as Research Fellow and University Lecturer. He also has ten years' experience as Head of IT Department in a Bank. His institutional address is Methodist University College Ghana, P.O. Box DC 940, Accra, and his research interest is in the effect of computerization on the Ghanaian society.

Dr. Kwadwo OWUSU is a Senior Lecturer at the Department of Geography and Resource Development, University of Ghana, Legon. His research interests are in the areas of climatology, climate impacts on agriculture variability and climate change adaptation. He is currently the Coordinator of the University of Ghana's Graduate program in Climate Change and Sustainable Development. *Department of Geography and Resource Development, University of Ghana, P.O. Box LG59, Legon. GHANA* Tel. +233 267528993. Email: kowusu@ug.edu.gh

Kenneth K. TSIVOR is currently an Instructor/Lecturer at Ghana Technology University College Accra. He is also a Ph.D. Fellow at Aalborg University, Copenhagen. Previously, he served as Senior Manager/Head Energy Systems, Ghana Telecom Training Centre and also as Project Counterpart, Vodafone Ghana (Ghana Telecom-Projects Department). He holds a Master Degree in Energy Engineering from South Bank University London. He holds a Post Graduate Certificate in Telecom Power Engineering from Nippon Telegraph and Telegraph Institute Osaka. He also holds a degree in Electrical Engineering from Kumasi Polytechnic, Ghana.

Musa USMAN is a lecturer at the Department of Mass Communication, University of Maiduguri, Nigeria. His teaching and research interest is International Broadcasting, Advertising, Communication Theory and Communication Research. He is a member of Faculty of Social Sciences, Board of Examiners, University of Maiduguri, Nigeria He has served as a resource person/Facilitator in several human resources capacity building workshops in Nigeria. He is a member of African Council for Communication Education. Email address: alhjmusa2@gmail.com. Tel: +2348067211251, +2348056173037

Joseph WILSON is a lecturer at the Department of Mass Communication, University of Maiduguri, Nigeria. His Teaching and research interest is in the field of Information and Communication Technology/New Media and Journalism. His most recent research (yet to be published) is on Online Interactivity in the Print Media in Nigeria. He is a Member of African Council for Communication Education (ACCE) and European Communication Research and Education Association (ECREA). He is also a member of Faculty of Social Sciences, Board of Examiners, University of Maiduguri, Nigeria. Email address: joeweee2003@yahoo.com, wilson@unimaid.edu.ng. Tel: +2348038399712,

Lightning Source UK Ltd.
Milton Keynes UK
UKOW05n1820310517
302420UK00012B/100/P

9 788793 102637